Springer Tracts in Civil Engineering

Series Editors

Sheng-Hong Chen, School of Water Resources and Hydropower Engineering, Wuhan University, Wuhan, China

Marco di Prisco, Politecnico di Milano, Milano, Italy

Ioannis Vayas, Institute of Steel Structures, National Technical University of Athens, Athens, Greece

Springer Tracts in Civil Engineering (STCE) publishes the latest developments in Civil Engineering - quickly, informally and in top quality. The series scope includes monographs, professional books, graduate textbooks and edited volumes, as well as outstanding PhD theses. Its goal is to cover all the main branches of civil engineering, both theoretical and applied, including:

- Construction and Structural Mechanics
- Building Materials
- Concrete, Steel and Timber Structures
- Geotechnical Engineering
- Earthquake Engineering
- Coastal Engineering; Ocean and Offshore Engineering
- Hydraulics, Hydrology and Water Resources Engineering
- Environmental Engineering and Sustainability
- Structural Health and Monitoring
- Surveying and Geographical Information Systems
- Heating, Ventilation and Air Conditioning (HVAC)
- Transportation and Traffic
- Risk Analysis
- Safety and Security

Indexed by Scopus

To submit a proposal or request further information, please contact:
Pierpaolo Riva at Pierpaolo.Riva@springer.com (Europe and Americas) Wayne Hu at wayne.hu@springer.com (China)

Marco Guerrieri

Fundamentals of Railway Design

 Springer

Marco Guerrieri
Department of Civil, Mechanical
and Environmental Engineering (DICAM)
University of Trento
Trento, Italy

ISSN 2366-259X   ISSN 2366-2603   (electronic)
Springer Tracts in Civil Engineering
ISBN 978-3-031-24032-4   ISBN 978-3-031-24030-0   (eBook)
https://doi.org/10.1007/978-3-031-24030-0

This Springer imprint is published by the registered company Springer Nature Switzerland AG
The registered company address is: Gewerbestrasse 11, 6330 Cham, Switzerland

Preface

This book offers a concise overview of the methods and criteria adopted for the design of railway infrastructures. Conventional railways are considered together with high-speed railways, tramways, metros, maglev, hyperloop systems, people movers, monorails and rack railways. Every system of transport is described in its main technical characteristics, capacities and construction costs. It is an introductory book to specific topics of the railway engineering field, and thus, the mathematical treatment is purposely brief and simplified.

The book is organized as follows:

Chapter 1 deals with the main descriptive models of train resistance which can be interestingly applied in designing railway lines as well as managing and regulating traffic flows.

Chapter 2 describes the various aspects influencing the (vertical and horizontal) alignment of ordinary and high-speed railways.

Chapter 3 deals with the principles involved in the construction of ballasted (conventional) tracks and ballastless (slab) tracks.

Chapter 4 deals with the analysis of the contact area between wheel and rail and the pressure distribution obtained by applying Hertz's theory. Since railway accidents result in a heavy loss of life and property damage, the chapter also analyses the derailment risk levels according to Nadal's formula.

Chapter 5 briefly describes the criteria for calculating ballasted tracks.

Chapter 6 presents the technique for analysing the track efficiency and especially the conformity of several geometric parameters of the track to normative threshold values.

Chapter 7 presents the main characteristics and classifications of switches and crossings.

Chapter 8 describes railway line configurations (i.e. single-track, double-track, triple-track and quadruple-track lines) and railway station types (e.g. wayside stations, junctions, terminals and seaport stations).

Chapter 9 classifies and briefly describes the main types of bridges which are commonly used in ordinary and high-speed railway lines.

Chapter 10 illustrates some techniques for tunnel design starting from the rock mass classifications.

Chapter 11 presents some traffic management systems (TMSs) and describes some models for evaluating station and line capacity with automated block and mobile block systems.

Chapter 12 briefly describes the technical characteristics of high-speed railways, the Transrapid and Hyperloop systems.

Chapter 13 illustrates the technical characteristics of heavy and light metros.

Chapter 14 deals with the main technical characteristics of tramways with conventional and ground-level power supply systems.

Finally, Chap. 15 deals briefly with the key technical characteristics of people movers, monorails and rack railways.

The book is addressed to civil engineering students, young engineers working in the field of railway design, as well as to engineers unfamiliar with railway engineering topics.

This book first appeared in Italian in 2017 under the title *Infrastrutture ferroviarie, metropolitane, tranviarie e per ferrovie speciali—Elementi di pianificazione e di progettazione*: this is the English version, revised and expanded. Special thanks are due to Giuseppina Zummo for her professional competence and accuracy in the English translation.

Palermo, Italy   Marco Guerrieri

Contents

Chapter 1
Train Resistance and Braking Distance

Abstract This chapter deals with the main descriptive models of train resistance which can be interestingly applied in designing railway lines as well as managing and regulating traffic flows.

1.1 Adhesion Force

Railway vehicle wheels are divided into [1, 2]:

- trailer wheels, unconnected to the engine;
- driving wheels, directly connected to the engine. Driving wheels transmit force to the rails, thus transforming torque into tractive force and causing the train to move.

The total weight on all driving wheels is termed adhesive weight (W_a). A railway vehicle transmits the rails a system of forces which, if summed up in vertical direction, correspond to the total weight of the vehicle ($W = \sum W_i$, see Fig. 1.1). Moreover, the driving wheels transmit the rails a system of horizontal forces resulting in the so-called tractive force F ($F = \sum F_i$, see Fig. 1.1).

In the direction of the movement, there are two different types of forces:

- active forces, whose sum is the tractive force F (depending on the speed, $F = F(v)$, and typical of any type of train);
- passive forces, or resistances to movement, whose sum is denoted with R.

If, at a first approximation, the train is considered as a material point of mass M and speed v, Eq. (1.1) is applied to describe, by means of appropriate specifications, all the movement phases of a railway vehicle from a mathematical point of view.

$$F = R + M \cdot \frac{dv}{dt} \leq F^* \tag{1.1}$$

$$F^* = f \cdot W_a \tag{1.2}$$

© The Author(s), under exclusive license to Springer Nature Switzerland AG 2023
M. Guerrieri, *Fundamentals of Railway Design*, Springer Tracts in Civil Engineering,
https://doi.org/10.1007/978-3-031-24030-0_1

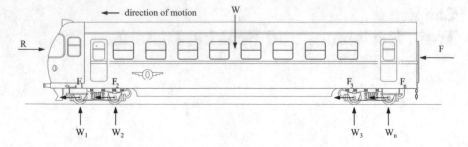

Fig. 1.1 Active and passive forces on a vehicle

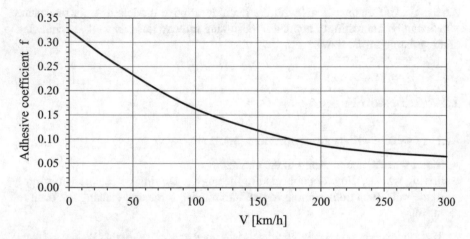

Fig. 1.2 Wheel-rail adhesive coefficient for speed values up to 300 km/h

where F* is the available adhesion force (acting on the wheel and rail contact point with a different direction depending on which phase the vehicle is in, either acceleration or braking), f is the wheel-rail adhesive coefficient and W_a the adhesive weight.

The coefficient f varies with speed and environmental conditions (e.g. dry, wet, dirty rail). A typical trend of the adhesive coefficient when the speed varies (in the absence of macroscopic sliding) is represented in Fig. 1.2.

Adhesive coefficient values on varying speed (in km/h) can be worked out through the following expression[1]:

$$f = 0.3216 - 0.0019 \cdot V + 3 \cdot 10^{-6} \cdot V^2 \tag{1.3}$$

---

[1] The expression is similar to Lamm et al.'s, used in the road sector [3]: $f = 0.59 - 4.85 \cdot 10^{-3} \cdot V + 1.51 \cdot 10^{-5} \cdot V^2$

By comparing the latter with Eq. (1.3) it is perfectly clear that, at the same speed, the adhesive-on-iron coefficient values are always lower than the corresponding on-road pavement values (asphalt pavements provide f values up to 0.85 with dry surface and to 0.50 with wet surface [3]).

Table 1.1 Adhesive coefficient values at a 50 km/h speed

Types of traction	Adhesive coefficient f
Electric traction with coupled axes	0.25
Electric traction with free axes	0.20
Diesel traction with coupled axes	0.20
Diesel traction with free axes	0.167
Steam traction	0.167

The values in Table 1.1 inferred from [2] concern dry, cleaned or sandy rails, for vehicle speeds below or equal to 50 km/h. Such values are reduced by 20% in foggy or rainy conditions and by 50% in case of greased or muddy rail rolling surfaces.

For speeds up to 300 km/h Eqs. (1.4) or (1.5)—the latter being adopted by the Spanish RENFE—can be also used [4]:

$$f = \frac{f_0}{1 + 0.01 \cdot V} \tag{1.4}$$

$$f = f_0 \cdot (0.2115 + \frac{33}{42 + V}) \tag{1.5}$$

where f_0 is the adhesive coefficient value for speeds near zero and V is the speed in km/h. The f_0 values are shown in Table 1.2.

The basic phases of the movement are easily obtained from Eq. (1.1):

- start-up phase: $F > R$, therefore $\frac{dv}{dt} > 0$ (accelerated movement);
- steady-state phase: $F = R$, therefore $\frac{dv}{dt} = 0$ (constant speed movement);
- inertia driving phase: $F = 0, R > 0, \frac{dv}{dt} < 0$ (decelerated movement due to passive resistance);

Table 1.2 Adhesive coefficient values for speeds near zero

f_0 coefficient		
SNCF (France)	Electric traction	0.33 –0.35
DB (Germany)	Diesel traction	0.30
	Electric traction	0.33
RENFE (Spain)	Diesel traction	0.22 – 0.29
	Electric traction (old-generation trains)	0.27
	Electric traction (new-generation trains)	0.31
USA	SD75MAC electric and diesel tractions	0.45

- braking phase: $F = 0$, $R > 0$, plus a braking force Q_T also acting on the vehicle (resulting in $0 = R + Q_T + M \cdot \frac{dv}{dt}$), $\frac{dv}{dt} < 0$ (with higher-intensity deceleration than inertia driving phase).

1.2 Resistances to Movement

Resistances to movement, whose sum corresponds to the term R in Eq. (1.1), are traditionally divided into two distinct categories:

- ordinary resistances, which develop in all the movement phases, independently of the road geometry (in that they do not depend on either the curvature or the gradient). The ordinary resistances are:

(1) the rolling resistance R_1 equal to the resistance sum of the kinematic pair spindle—bearing (R_1') and the pair wheel—rail (R_1''). For the latter, the rate imputable to the rolling friction can be estimated with the relation [1]: $R_{1,v}'' = P\sqrt{\frac{2 \cdot d}{R_a}}$, being P the weight on the wheel, R_a its radius and d the depth of the wheel-induced vertical deformation of the top rail surface ($d \approx 18 \cdot 10^{-8}$ m).

In terms of unitary resistance (resistance/weight ratio: $r = \frac{R}{P}$), the following can be assumed as indicative values: $r_1' = 4.5$ daN/t for the start-up phase; $r_1' = 1.8$ daN/t at 90 km/h; $r_{1,v}'' = 0.5 - 1$ daN/t.

(2) air resistance R_2 due to the system of forces displaying during the motion of a body within a fluid (air flow). Such a resistance results from the sum of the overpressure on the vehicle front part, the lateral side resistance and the vehicle back turbulences. In completely general terms, for speeds up to 300 km/h, Eq. (1.6) can be considered, at a first approximation, valid as it takes only the first contribution (front part overpressure) in consideration:

$$R_2 = \rho \cdot S \cdot c \cdot V^2 \tag{1.6}$$

With ρ, S, c and V denoting, respectively, the air density, the frontal cross-sectional area, the aerodynamic drag coefficient and the vehicle speed.

- Accidental resistances, that is, due to railway alignment:

    (a) grade resistance (R_i) to the movement on an inclined track (uphill motion);
    (b) curve resistance (R_c).

1.2.1 Resistance on Horizontal and Straight Tracks

Resistance on horizontal and straight railway segments (R_0) is given by the sum of the rolling resistance (R_1) and the air resistance (R_2). In general, the relations connecting the specific resistance ($r_0 = R_0/P$) to speed, both deduced experimentally, are binomial or trinomial:

$$r_0 = a + b \cdot V^2 \tag{1.7}$$

$$r_0 = a + b \cdot V + c \cdot V^2 \tag{1.8}$$

The formulas used to estimate r_0 consider the contribution given by locomotives, coaches and wagons to resistance. The following expressions (cf. Fig. 1.3) are considered to be valid in open-sky railway sections (i.e. embankment, cutting and bridge sections) [2], in which V is in km/h and r_0 in daN/t:

- Clark's formula (valid for low speeds):

$$r_0 = 2.4 + \frac{V^2}{1000} \tag{1.9}$$

- Erfurt's formula (valid for average speeds):

$$r_0 = 2.4 + \frac{V^2}{1300} \tag{1.10}$$

- Von Borries formula (valid for high speeds):

$$r_0 = 1.6 + 0.3 \cdot V \cdot \frac{V + 50}{1000} \tag{1.11}$$

- Barbier's formula (valid for high speeds)

$$r_0 = 1.6 + 0.456 \cdot V \cdot \frac{V + 10}{1000} \tag{1.12}$$

The relations more recently adopted by the Italian infrastructure manager of the railway network (RFI) are binomial [5]:

- for passenger trains

$$r_0 = 1.94 + 2.65 \cdot \left(\frac{V}{100}\right)^2 \tag{1.13}$$

- for freight trains

$$r_0 = 2.04 + 5.01 \cdot \left(\frac{V}{100}\right)^2 \tag{1.14}$$

Equations (1.13) and (1.14) overestimate the resistance values for new-generation trains (in which the resistances of the pair spindle—bearing, R_1', are lower than old-generation vehicles) and for the railway track equipped with long welded rails (LWR).

In France the SNCF (Société Nationale des Chemins de fer Français) adopts the following Eq. (1.15) for the high-speed railway network TGV (Train à Grande Vitesse) [5]:

$$r_0 = \left(1.3 \cdot \sqrt{\frac{10}{p}} + 0.01 \cdot V\right) \cdot P + \left(0.0021 \cdot S + 0.025 \cdot p_e \cdot \frac{L-45}{100} + \sum k_i\right) \cdot V^2$$

(1.15)

in which:

- r_0 is the specific resistance [daN/t];
- P is the train weight [t];
- p is the axial weight [t];
- S is the frontal cross-sectional area of the train [m^2];
- L is the total train length [m];
- p_e is the partial perimeter of the train from rail to rail [m];
- $\sum k_i$ is the sum of correction coefficients for the fairing defects (e.g. no fairing in bogies, protuberances, aerodynamic limitations in vehicle front and back, etc.).

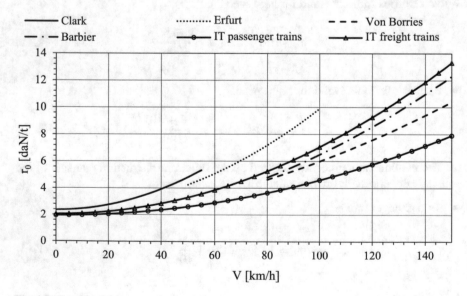

Fig. 1.3 Specific resistance on horizontal and straight railway segments

An experimental study carried out in France [6] showed that for a train travelling at the speed of 260 km/h, 80% of the aerodynamic resistance is due to its body, 17% to its contact points with the pantograph, 3% to vehicle brakes (discs) and other additions.

Researches carried out on the Shinkansen high-speed train network in Japan have led to the following formula [7]:

$$R_0 = (1.2 + 0.022 \cdot V) \cdot P + (0.013 + 0.00029 \cdot l) \cdot V^2 \tag{1.16}$$

where R_0 is expressed in daN, P denotes the total train weight [t], V the speed [km/h] and l the total train length [m].

The specific resistance on railway tunnel segment $r_{0,G}$, on equal terms (speed, train type, etc.), is systematically higher than that in open-sky sections (i.e. embankment, cutting and bridge sections) $r_{0,A}$. It thus results: $\frac{r_{0,G}}{r_{0,A}} = 1.7-1.8$.

Such an increase in resistance cannot be neglected on lines with a considerable tunnel extension (e.g. underground lines) and/or with very long single tunnels,[2] also on account of the ever-growing train speed on high-speed railway networks[3] [8]. In these cases, in order to estimate only the aerodynamic resistance, it is convenient to apply specific analytic expressions like the following one which takes the effective train length and lateral friction coefficients into account [7]:

$$R_a = \frac{1}{2} \rho A' V^2 (C_{dp} + \frac{\lambda'}{d'} \cdot l) \tag{1.17}$$

where:

- V stands for the train speed;
- ρ is the air density;
- A' stands for the frontal cross-sectional area of the train;
- C_{dp} is the coefficient of the pressure drag caused by the train fore-and after-bodies;
- d' is the hydraulic diameter of the train;
- l stands for the train length;
- λ' is the hydraulic friction coefficient caused by the connecting parts between trains, photographs, structures under the train, etc.;

By way of example Table 1.3 shows the coefficient values of the relation (1.17) obtained from the Japanese high-speed railway lines [7, 9].

If the characteristics of the tunnel wall surfaces are known, the aerodynamic resistance can be assessed with Parshall and Hobart's relation [10]:

---

[2] Some remarkable examples are the 57 km long tunnel planned on the line Turin–Lyon, the 53.85 km long Seikan Tunnel in Japan, the over-50 km long Channel Tunnel etc.

[3] The TGV in France set the record speed of 574.8 km/h; the same TGV technology in South Korea reached the speed of 330 km/h. The Eurostar in England set a national record of 334.7 km/h, while the Japanese Shinkansen reaches an operational speed of 320 km/h.

Table 1.3 Coefficient values in Eq. (1.17)

Train series denomination	Cross section area A′ [m²]	Hydraulic diameter d′	Hydraulic friction coefficient λ′	Coefficient C_{dp}
0	12.6	3.54	0.017	0.20
200	13.3	3.64	0.016	0.20
100	12.6	3.54	0.016	0.15

Table 1.4 Values of coefficient γ in Parshall and Hobart's relation [10]

Tunnel type	Coefficient γ
Smooth-walled tunnel	0.020
Rough-walled tunnel	0.027
Masonry vaulted tunnel	0.02 –0.04
Rock-walled tunnel	0.06 –0.08
Tunnel with wooden armour	0.10 –0.15

$$R_2 = \gamma \cdot \frac{L \cdot V^2}{2 \cdot d \cdot g \cdot \omega} \tag{1.18}$$

in which:

- d is the mean tunnel diameter $d = \frac{4A}{P_{er}}$, where A is the tunnel area and P_{er} its perimeter;
- L is the tunnel length;
- ω is the specific air volume (0.752 m³/daN for T = 15° and atmospheric pressure);
- γ is a coefficient depending on the type of tunnel wall surface (see Table 1.4).

Experimentations carried out in Italy in the 1980s in some tunnels of the Florence–Rome high-speed railway line with different lengths and a 53.5 m² cross section area confirmed, however, the validity of the binomial relation (Eq. (1.7)) for a speed up to 250 km/h.

The values of coefficients a and b of Eq. (1.7) are given in Table 1.5 [5].

1.2.2 Resistance Due to Gradient

The resistance due to gradient, also named "grade resistance", develops during the movement on a line section with a non-null gradient (inclined rolling plane). By denoting the rail inclination angle with α, with respect to the horizontal (Fig. 1.4), and recalling that for very small angles, as those typical in railway lines, senα ≈ tgα, the resistance can be estimated through the expression:

Table 1.5 Values of coefficients a and b in Eq. (1.7)

Train	Resistance to movement $r_0 = a + bV^2$ [daN/t]			
	Open-sky sections (embankment, cutting and bridge sections)		Tunnel sections	
	a	b	a	b
2 Ale 601 + trailer	0.99684	0.00025	0.62017	0.00046
Locomotive + 4 coaches	1.3845	0.00021	1.60348	0.00035
Freight train DB (Deutsche Bahn)	1.6952	0.00027	1.24463	0.00051
ETR (IT Rapid Electric Train) 500	0.90404	0.00012	0.833	0.00021

$$R_i = P \cdot i \cdot \frac{1}{\sqrt{1 + i^2}} \tag{1.19}$$

where:

- P indicates the train weight;
- i denotes the gradient slope: $i = tg\,(\alpha)$.

Considering that i^2 can be neglected with respect to unity, the relation (1.19) can be approximated with Eq. (1.20):

$$R_i = P \cdot i \tag{1.20}$$

It follows that the specific grade resistance r_i expressed in daN/t coincides numerically with the value of the gradient slope expressed in per mille:

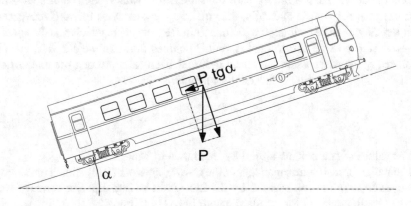

Fig. 1.4 Grade resistance

Table 1.6 Coefficients in Röckl's formula

Radius R [m]	k_1	k_2
≥ 350	650	55
350–250	650	65
250–150	650	30

$$r_i = i \tag{1.21}$$

For instance, a train travelling on a track with a 2‰ gradient is subject to a specific resistance $r_i = 2$ daN/t.

If the vehicle runs along the gradient downhill, the resistance R_i becomes a full-fledged traction force. Thus, in this specific case, it is computed as a negative resistance. In the sections in which the part of the train l_1 is on the slope gradient i_1 and the other part l_2 is on the slope gradient i_2 (with $l_1 + l_2 = l$ corresponding to the total train length), neglecting the effect of the vertical curve between the two gradients, the specific resistance can be estimated with the expression:

$$r_i = \frac{l_1 \cdot i_1 + l_2 \cdot i_2}{l_1 + l_2} \tag{1.22}$$

1.2.3   Resistance Due to Curvature

During the movement in curve, the vehicle meets a resistance due to the rigid fit between wheels and axle (wheelset) and to parallel axes in the same bogie; for this reason, despite the characteristic truncated cone profile of wheel treads and the inclination of the rail heads (generally with 1:20 gradient, see Chap. 2), the wheel outside the curve traces a longer path than the inside wheel. This causes slips and collision between wheels and rails, thus offering resistance to forward movement. Such resistance can be estimated with the relation $R_c = r_c \cdot P$, in which r_c is the specific curve resistance expressed in daN /t, and P denotes the train weight. The specific resistance r_c increases when the curve radius R decreases. Among the expressions mostly used for estimating r_c is Röckl's formula:

$$r_c = \frac{k_1}{R - k_2} \tag{1.23}$$

The values of coefficients k_1 and k_2 are shown in Table 1.6.

The Italian infrastructure manager of the railway network (RFI) adopts the average unit resistance values represented by the histogram in Fig. 1.5. Such values were obtained by Bauman [1] for standard gauge lines (1435 mm).

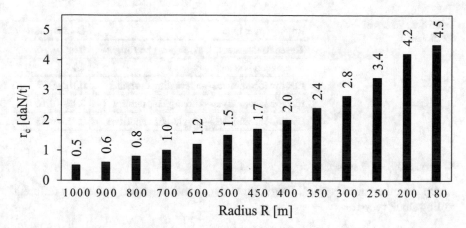

Fig. 1.5 Specific curve resistance

The specific curve resistance is negligible in standard lines and radiuses R > 1000 m. Röckl's relations can be applied also to narrow-gauge lines:

- For 1 m gauge:

$$r_c = \frac{100}{R-20} \quad [\text{daN/t}] \tag{1.24}$$

- For 0.75 m gauge:

$$r_c = \frac{300}{R-10} \quad [\text{daN/t}] \tag{1.25}$$

If the curve length S_v is lower than the train length l, the corresponding specific curve resistance $r_c{}'$ can be calculated with the expression $r_c' = r_c \cdot \frac{S_v}{l}$.

1.2.4 Resistance Due to Inertia

The inertia resistance R_{in} occurs during both the acceleration and deceleration phases of motion.

Typical longitudinal acceleration (dv/dt) values for different rolling stock types are:

- freight trains: 0.2 –0.4 m/sec²;
- intercity trains: 0.4 –0.6 m/sec²;
- suburban trains: 0.6 –0.8 m/sec²;
- metros: 0.8 –1.0 m/sec²;

Table 1.7 Values of coefficient μ

Rolling stock type	Coefficient μ
Towed rolling stock (passengers and freight)	0.08 – 0.07
Steam locomotives	0.15
Electric locomotives—alternating current	0.18 – 0.20
Electric locomotives—three-phase current	0.13 – 0.16
Electric locomotives—single-phase current	0.35 – 0.45

- trams: 0.8 –1.2 m/sec^2.

On the other hand, typical longitudinal deceleration (dv/dt) values for different rolling stock types are:

- conventional freight trains: 0.10 m/sec^2;
- express freight trains: 0.25 m/sec^2;
- passenger trains: 0.40 –0.50 m/sec^2;
- suburban railways and metros: 0.60 m/sec^2.

By particularising Eq. (1.1), it follows:

$$R_{in} = F - R = \frac{P}{g} \cdot \frac{dv}{dt} \tag{1.26}$$

It is clear from Eq. (1.26) that in the stages of non-uniform motion, being $\frac{dv}{dt} \neq 0$, the tractive force F (engine effort) must overcome R_{in} besides the other resistances due to movement.

In order to consider that a train is not a rigid body but rather involves masses with their own rotational motion and their own inertial properties, the total vehicle mass is amplified with a factor $(1 + \mu)$:

$$R_{in} = (1 + \mu) \cdot \frac{P}{g} \cdot \frac{dv}{dt} \tag{1.27}$$

The characteristic μ values are shown in Table 1.7 [2].

The $(1 + \mu)$ value of a train composed of n individual rolling stocks (i.e. locomotives, coaches, wagons, etc.), each having weight P_i and inertia amplification factor $(1 + \mu_i)$, can be calculated by the weighted mean:

$$(1 + \mu) = \frac{\sum_{i=1}^{n} (1 + \mu_i) \cdot P_i}{\sum_{i=1}^{n} P_i} \tag{1.28}$$

1.3 The Traction Force and Electric Traction Standards in Europe

Also the tractive force (or tractive effort) F in Eq. (1.1) depends on the speed v. The function linking F with v (F = F(v)) depends on the engine type and characteristics. Assumingly, in numerous technical applications the train engine gives, at any operating speed, a constant value of power N (ideal engine); in this case it yields:

$$N = F \cdot V = \text{const} \qquad (1.29)$$

Considering the efficiency of the transmission system η (with $\eta = 0.75 - 0.80$ for diesel locomotives and $\eta = 0.95 - 0.97$ for electric locomotives) and expressing the power in kW, it follows:

$$N = \frac{F \cdot v}{367 \cdot \eta} \qquad (1.30)$$

Once the power value N of an ideal engine is known, the relation F = F(v) can be inferred easily. By using this equation and estimating the laws of resistance to motion, Eq. (1.1) can be particularised in order to then examine the train motion phases (start-up, constant speed, braking, etc.) for a plethora of purposes (travelling time estimation, train scheduling, line capacity estimation, etc.).

The modern trains are powered electrically. The main tractive standards adopted in the European railways lines are as follows:

- direct current systems:

  - 3000 V in Italy, Czech Republic, Poland, Slovenia, Spain;
  - 1500 V mainly in France;
  - 750 V through the third track (especially in the south of Great Britain);

- single-phase alternating current systems:

  - 15,000 V 16 2/3 Hz in Austria, Germany, Switzerland;
  - 25,000 V 50 Hz on all the new European high-velocity lines;

- three-phase alternating current systems:

  - 3600 V 16 2/3 Hz (Central Europe and Italy until 1976).

1.4 Line Performance Levels

In order to characterise the unevenness of a railway line circuit, a global plano-altimetric specific resistance, named "compensated gradient" i_c (or else, fictitious

Table 1.8 Railway line performance levels established by the Italian infrastructure manager of the railway network (RFI)

Performance level	i_c [daN/t]	Performance level	i_c [daN/t]	Performance level	i_c [daN/t]
1	4.5	12	12.0	23	24.6
2	5.0	13	12.9	24	25.7
3	5.5	14	13.8	25	27.8
4	6.0	15	14.6	26	29.3
5	6.5	16	15.8	27	30.8
6	7.0	17	17.0	28	32.5
7	7.7	18	18.4	29	34.2
8	8.4	19	19.8	30	37.5
9	9.2	20	20.9	31	40.5
10	10.0	21	21.9	–	–
11	11.0	22	22.7	–	–

gradient) is applied conventionally and derived from the sum of the specific grade resistance r_i and the specific curve resistance r_c:

$$i_c = r_i + r_c \tag{1.31}$$

For example, the Italian railway network is subdivided into basic sections (generally longer than 2 km), called "loading sections", each with a constant fictitious gradient $i_c = const$.

In each loading section, the specific grade resistance and the specific curve resistance can vary but their sum always results in the same i_c value. Equation (1.31) makes it clear that the compensated gradient of the same basic section of a railway line has a different value in the two directions of travel. The Italian infrastructure manager of the railway network (RFI) defined the performance levels as shown in Table 1.8.

Should there be, inside the loading section, a few-hectometre-long stretch with a compensated gradient higher than the performance level characterising the whole section, an additional performance level, associated to the main one, is used. For instance, a section with index 11_6 identifies a loading section which has the performance level 11.0 daN/t and a short stretch inside with a compensated gradient below 7.0 daN/t.

1.5   Line Virtuality Levels

The virtual length of line L_v is estimated to provide a basis for comparison.

L_v corresponds to the length that a given railway line would have if it were horizontal and straight ($r_c = 0$ daN/t and $r_i = 0$ daN/t) in order to be equivalent to its real length L_r, with respect to a particular selected parameter.

In terms of energy, the virtual length is equal to the length of an ideal railway line, with a horizontal and straight alignment, which determines the same work as the traction force on the real line. Therefore, if the real railway alignment extends on only one grade and its curvature ($\rho = \frac{1}{R}$) is constant and different from zero, this equation is valid:

$$L_v \cdot 1000 \cdot P \cdot r_0 = L_r \cdot 1000 \cdot P \cdot (r_0 + r_i + r_c) \tag{1.32}$$

where P is the train weight expressed in tonnes. From Eq. (1.32), it follows:

$$L_v = L_r \cdot \frac{r_0 + r_i + r_c}{r_0} = L_r \cdot (1 + \frac{r_i + r_c}{r_0}) \tag{1.33}$$

It should be also considered that, if the grade is travelled downhill, in Eq. (1.33) r_i has a negative value, thus making the virtuality factor $\left(1 + \frac{r_i + r_c}{r_0}\right)$ equal to, higher or lower than 1, and consequently making L_v equal to, higher or lower than L_r, once the direction of travel is established.

In general, the alignment of a railway line consists of n different sections, each with a given length l_j and resistances $r_{0,j}$, $r_{i,j}$, $r_{c,j}$; therefore, the virtual length is estimated with the following expression:

$$L_v = \sum_{j=1}^{n} l_j \cdot (1 + \frac{r_{i,j} + r_{c,j}}{r_{0,j}}) \tag{1.34}$$

The $r_{0,j}$ value is to be particularised in presence of tunnels in single sections, as previously explained.

The virtual line length can be also measured by using other parameters such as traction costs, transportation costs, operational costs, total travel time costs, etc. [2].

1.6 The Braking Distance

While braking, a generic wheelset of weight P_i is, by means of an appropriate braking system,[4] subject to a radial force Q_i generating a friction force Φ to rotation (Fig. 1.6) with the following value:

---

[4] Brake systems can be classified as follows: *shoe* or *block or tread brakes* (the friction force is generated on the wheels by the pressure of metal or synthetic shoes) and *disc brakes* (the braking force is obtained by friction on steel discs, or cast iron, fixed to the axle of wheelsets) [10].

The braking force can be transmitted by using the following systems: air braking, electropneumatic braking, electromagnetic braking, electrodynamic braking [10].

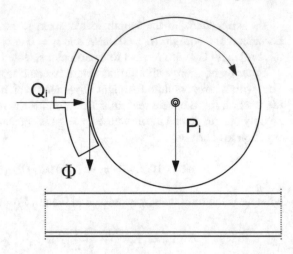

$$\Phi = f' \cdot Q_i \tag{1.35}$$

where f' is the friction factor between the brake shoe and the rim of the wheel.

In order to avoid wheel slipping (which would increase the braking distance and concomitantly would accelerate wheel and rail wear phenomena), the friction force Φ must be lower than the available adhesion; thus, the following condition needs to be satisfied (see Eq. (1.2)):

$$\Phi = f' \cdot Q_i \leq f \cdot P_i \tag{1.36}$$

At the limit of adhesion, it results $f' \cdot Q_i = f \cdot P_i$, and thus:

$$\frac{Q_i}{P_i} = \frac{f}{f'} \tag{1.37}$$

Therefore, at the limit of adhesion, the ratio between the force applied to the wheels by the braking system and the wheelset weight equals the ratio between the rim/rail friction factor f and the rim/brake shoe friction factor f'. Both friction factors being dependent on speed v (i.e. $f = f(v)$; $f' = f'(v)$), the ratio Q_i/P_i is also a function of it. It follows that for an efficient braking, at the limit of adhesion, the vehicle brake system should ideally assure a continuous variation of the ratio Q_i/P_i in function of v.

As a matter of fact, there are no brake systems satisfying the above requirement and indeed brake systems are used at one or more steps, that is, in order to provide only one ratio Q_i/P_i (value range 0.75 –0.85) or more Q_i/P_i ratios (value range 1.20 –1.60), each being typical of a given speed field.

In case of a single-step brake system, the most critical condition occurs near zero speed ($v = 0$) as follows:

$$\frac{f_{v=0}}{f'_{v=0}} \cong 0.7 \tag{1.38}$$

$$\frac{Q_i}{P_i} = \lambda_r \leq 0.7 \tag{1.39}$$

With $\lambda_r = \lambda_r(v)$ denoting the braked weight percentage of the wheelset.

In a train consisting of a wheelset number equal to i, the ratio between the total braking force Q_T and the total train weight P_T is:

$$\frac{Q_T}{P_T} = \frac{\sum_i f' \cdot Q_i}{P_T} = f' \cdot \sum_i \frac{Q_i}{P_i} \cdot \frac{P_i}{P_T} = f' \cdot \lambda_r \frac{1}{P_T} \cdot \sum_i P_i = f' \cdot \lambda_r \tag{1.40}$$

In short, λ_r also corresponds to the braked weight percentage of the train. By denoting with λ_c a coefficient named conventional braked weight percentage of a train, with a unit value for $\lambda_r = 0.7$, the previous expression can be written as:

$$\frac{Q_T}{P_T} = f' \cdot 0.7 \cdot \lambda_c \tag{1.41}$$

All this said, the braking distance (s) with emergency braking[5] (or braking "space"), corresponding to the distance necessary to stop the vehicle from the generic speed v_i, can be calculated with Eq. (1.1), by imposing the condition $F = 0$, introducing Q_T among resistances to the movement and considering that, before applying the braking force—starting from the theoretical instant when the braking should start—a delay[6] t_p is inevitable. During the driver's perception/reaction time delay t_p the train travels the distance $s_p = v_i \cdot t_p$.

And finally, once acceleration is defined as $a = \frac{dv}{dt} = \frac{v \cdot dv}{ds}$, the following relation can be easily shown to calculate the space $s(v_i)$ required to stop the train from the initial speed $v = v_i$ to the final speed $v_f = 0$:

$$s(v_i) = v_i \cdot t_p + \frac{(1 + \mu)}{g} \cdot \int_{v_i}^{0} \frac{v}{(0.7 \cdot \lambda_c \cdot f') + (r_1 + r_c + \frac{KSv^2}{P_T} \pm i)} \cdot dv \tag{1.42}$$

---

[5] The braking distance is the key kinematic parameter for regulating train traffic (due to its vital importance in terms of safety) and, indirectly, it also affects the line capacity.

[6] t_p corresponds to the driver's perception/reaction time during the visual driving mode in railway and tramway lines.

Table 1.9 Coefficient φ values in pedelucq's formula

V [km/h]	70	75	80	85	90	95	100
φ	0.06099	0.06216	0.06257	0.06308	0.06346	0.06407	0.06470
V [km/h]	105	110	115	120	125	130	135
φ	0.06560	0.06668	0.06802	0.06952	0.07096	0.07206	0.07215
V [km/h]	140	145	150	155	160	165	170
φ	0.07303	0.07348	0.07418	0.07482	0.07547	0.07645	0.07745
V [km/h]	175	180	185	190	195	200	–
φ	0.07816	0.07900	0.07998	0.08096	0.08197	0.08296	–

Table 1.10 Characteristic values of the braked weight percentage

Train Type	MA80	MA90	MA100	ME100	ME120	V120	V140	V160	TGV V160
λ_c	0.47	0.5	0.57	0.60	0.77	0.87	0.97	1.25	1.25

Generally, in order to determine the braking distance, the empirical relations specified in technical regulations[7] or in scientific literature are used, e.g. Bricka's, Maison's [10] or Pedelucq's formulas [11]. The last is shown below:

$$s = \frac{V^2}{\frac{1.09375 \cdot \lambda_c}{\phi(V)} + \frac{0.127}{\phi(V)}} \pm 0.235 \cdot i \tag{1.43}$$

where V is expressed in km/h, λ_c is expressed in decimal notation (e.g. for a braked weight percentage 110%, λ_c is 1.1) while the coefficient φ, obtained experimentally, assumes the values shown in Table 1.9.

The characteristic values of the conventional braked weight percentage of some train types are shown in Table 1.10 [12].

Figure 1.7 shows the diagram of braking distances in function of the braked weight percentage and speed values.

Due to the high braking distance values, train protection from obstacles on the track requires the use of driver's warning systems by means of signals and alarms.

The braked weight percentage is usually determined in every vehicle. The braked weight Q_T can be inferred indirectly by standardised brake tests: the first step is to measure experimentally the distance "s" which the vehicle requires to stop, starting from a given vi value (e.g. UIC, Union Internationale des Chemins de fer sets V

---

[7] See, for instance, the braking model describes in the regulation "SCMT—Italian Train Protection System" published by Italian infrastructure manager of the railway network (RFI).
(Cod. RFI TC.PATC SR CM 03 M59 C).

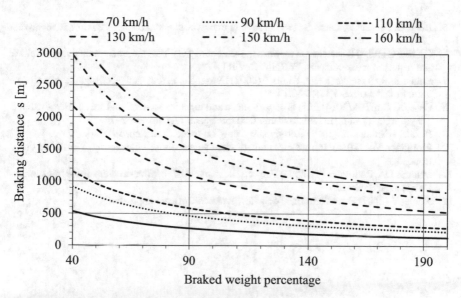

Fig. 1.7 Breaking distance in function of the braked weight percentage (λ_c) and speed on a horizontal alignment (i = 0)

= 120 km/h). From Eq. (1.43) λ_c is obtained and if the total vehicle weight P_T is known, Q_T can be calculated. Considering the estimation procedure, the obtained value is a conventional measure expressing the vehicle braking capacity. P_T and Q_T are stamped into the exterior fairings of coaches, wagons, etc. All this said, it is vital for the head locomotive to brake less heavily than the hauled coaches and wagons in order that the latter cannot exert a push on the former during the braking stages. Therefore, in composing trains an appropriate ratio is considered between the locomotive and vehicle braked weight percentages. Every line (or its section) is distinguished by a *braking level* which allows to establish the attainable maximum speed of a certain train in function of its braked weight percentage. For example, in Italy there have been identified 10 braking levels (I_a I, II, III...IX) which increase when the line gradient "i" increases (e.g. I_a: 0‰ < i ≤ 4‰; IX: 30‰ < i ≤ 35‰).

References

1. Nicolardi A (1956) Special railways (in Italian, *Ferrovie speciali*), Casa editrice Dott. Carlo Cya
2. Vicuna G (1968) Railway engineering: organizations and techniques (in Italian, *Organizzazione e tecnica ferroviaria*), CIFI
3. Mauro R (2003) Highway engineering (in Italian, *La geometria stradale*), Hevelius edizioni
4. Lozano JA, Juan de Dios Sanz JF, Mera JM (2012) Railway traction. Reliability and safety in railway. InTech

5. Bono G, Focacci C, Lanni S (1997) Railway track (in Italian, *La sovrastruttura ferroviaria*), CIFI
6. Guihew C (1983) Resistance to forward movement of TGV-PSE trainsets: evaluation of studies and results of measurements. Fr Railw Rev 1(1)
7. Raghu S, Raghunathana HD, Kimb T (2002) Setoguchi. aerodynamics of high-speed railway train. Prog Aerosp Sci 38:469–514
8. Choi JK, Kim HK (2014) Effects of nose shape and tunnel cross-sectional area on aerodynamic drag of train traveling in tunnels. Tunn Undergr Space Technol 41:62–73
9. Hara T, Nishimura B (1967) Aerodynamic drag on train. RTRI Rep, 591
10. Profillidis VA (2022) Railway planning, management, and engineering. (Fifth Edition), Routledge
11. Mantras DA, Rodriguez PL (2003) Ferrocarriles: ingeniería e infraestructura de los transportes. Universidad de Oviedo
12. Noblet Y (2007) Formulaire tracé de voie, divers. 2007 (teaching material)

Chapter 2
The Alignment Design of Ordinary and High-Speed Railways

Abstract In railway systems the alignment refers to the centre line of the railway track. The alignment of ordinary and high-speed railways is composed of vertical and horizontal elements. The vertical alignment consists of straight (tangent) railway grades and the circular or parabolic curves that connect these grades. The horizontal alignment includes the straight (tangent) sections of the railway, the circular curves and the transition curves. The various aspects influencing alignment are discussed in this chapter.

2.1 Track Gauge

The track gauge is the distance between the inner edges of the heads of rails in a track, normally measured at 14 mm below the top surface of the rail.

Every country utilises one or more track gauge types, ranging from the minimum value of 600 mm to the maximum value of 1676 mm (see Table 2.1). Today the most adopted value is 1435 mm and is called *standard gauge*. In several countries, there are still broad-gauge and narrow-gauge lines in operation. However, terms such as broad-gauge and narrow-gauge do not have any fixed meaning beyond being materially wider or narrower than the standard gauge. Standard gauge railway lines have the advantage of greater traffic capacity, speed and safety with respect to narrow-gauge railway lines.

In order to inscribe the rolling stock easily, to minimise the noise and especially to reduce the rolling resistance, the gauge is appropriately increased in the horizontal curves with a smaller radius than 275 m.

The railway wheels have a truncated cone profile with 1/20 inclination (Fig. 2.1). When the gauge in curve increases, it allows the external wheel to roll on a rolling rim with a diameter larger than that in the internal wheel, thus getting the former to travel a bigger distance. Imposing that this longer travelling coincides numerically with the greater length in the external rail compared to the internal one (condition which theoretically avoids wheel sliding), it results that the gauge increase Δs, expressed in mm, is:

Table 2.1 Gauge type [1, 2]

Gauge type	Dimension [mm]	Main countries where applied
Standard (type 1)	1435	UK, USA, Canada, Persia, China, Italy, etc
Broad (type 2)	1168/1674	Portugal, Spain
Broad (type 3)	1676	India, Pakistan, Brazil, Argentina
Broad (type 4)	1520/24	Russia, Finland
Narrow (type 5)	1067	Africa, Japan, Australia, New Zealand
Narrow (type 6)	1065	Portugal
Narrow (type 7)	1000	India, France, Switzerland, Argentina
Narrow (type 8)	950	Italy
Narrow (type 9)	760	Austria, Former Yugoslavian countries
Narrow (type 10)	700	Argentina, Denmark, Indonesia, Spain etc

$$\Delta s = \frac{10 \cdot d_c \cdot D}{R} - J_s \qquad (2.1)$$

where:

- d_c is the distance between the wheel-rail contact points of the two rails, equal to around 1500 mm;
- D is the diameter of the wheel rolling circle, corresponding to the position of the wheelset centred on the railway track;
- R is the radius of the horizontal curve;
- J_s is the clearance, on a straight section, between the wheel flange and the gauge face of the rail.

The Italian infrastructure manager of the railway network (RFI) adopts the values shown in Table 2.2, lower than the theoretical ones which can be calculated with Eq. (2.1); this clearly implies some sliding, even though minimum.

The enlargement is obtained by modifying the planimetric position of the rail inside the curve so as to deviate the rail by 1 mm/m; on the other hand, the outside rail must continue to have an alignment parallel to the axis of the track, so as to keep to its guiding function of the wheel flange. The enlargement must be located:

- on the transition curve (parabola or clothoid): along its development, so that the enlargement is totally completed in the tangent point between the transition curve and circular curve;
- with no transition curve (parabola or clothoid): on the straight line, with the enlargement completed in the tangent point of the straight line and circular curve.

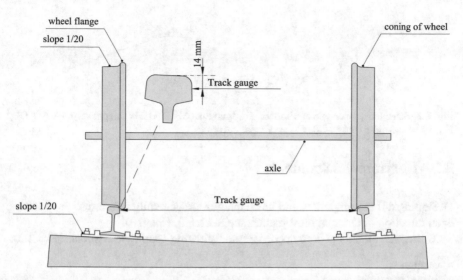

Fig. 2.1 Track gauge

Table 2.2 Gauge values in curve

Curve radius R value		Gauge
From R [m]	To R [m]	[mm]
∞	275	1435
< 275	250	1440
< 250	225	1445
< 225	200	1450
< 200	175	1455
< 175	150	1460
< 150	–	1465

For railway tracks with PRC (precompressed reinforced concrete) sleepers, the Italian Railways set the following tolerance limits with regard to the values shown in Table 2.2:

- construction tolerance (maximum deviations from the theoretical value acceptable in new railway line constructions):

  - − 1 mm, + 2 mm for ordinary railway lines;
  - 0, + 2 mm for high-speed railway lines;
  - railways into operation: + 7 mm, − 2 mm, except for the sections with a 1465 mm gauge where the tolerance is + 5 mm, − 2 mm.

Finally, the difference in gauge values between one sleeper and the next must not overcome 1 mm.

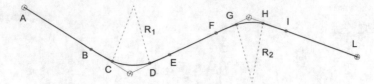

Fig. 2.2 Schematised example of a horizontal alignment (AB, EF, IL = straight sections, BC, DE, FG, HI = transition curves, CD, GH = circular curves)

2.2 Horizontal Alignment

Designing railway alignment and its chainages are generally referred to the track of even direction of the train movement.[1](see Sect. 8.3, Chap. 8).

Horizontal alignment is composed of the following geometric elements (Fig. 2.2):

- straights;
- circular curves;
- transition curves;
- polycentric and polycentric clothoidal (rarely used) curves.

The geometric element mainly affecting the maximum line speed is the circular curve radius. In Italy, RFI uses a minimum radius $R_{min} = 150$ m.

2.2.1 Straight Sections

Considering that railways are guided transport systems, no limit is set to the maximum length straight sections (also called tangent sections). Rather, straight sections should be designed as long as possible not only for providing users with greater comfort but also for reasons of functionality, route speed and safety.

On the contrary, it is necessary to ensure a minimum value for the straight length (L_{min}) in order to allow the vehicle body to restore the original vertical position, once the rolling stock has exited from the curve and entered the straight line. On an international level the below-listed minimum lengths have been identified [3]:

- Japan

$$L = \frac{1.5 \cdot V}{3.6} \quad [m] \tag{2.2}$$

---

[1] In several countries, railway directions are described as *even* and *odd*. The "even direction" is usually north- and eastbound, while the "odd direction" is south- and westbound. Trains travelling "even" and "odd" usually receive even and odd numbers as well as respective track and signal numbers.

This limit is in use for a tangent between reverse curves (i.e. when a curve to the left or right is followed immediately by a curve in the opposite direction) and considers that the vehicle body takes around 1.5 s to restore the original vertical position after coming out of the curve and entering the straight line [3].

- France (SNCF)

$$L_{min} = 30\,m \tag{2.3}$$

- Italy (RFI)

$$L_{min} = 30\,m \tag{2.4}$$

for speeds between 100 and 160 km/h, for both reverse and broken back curves (i.e. two curves of the same direction joined by a short straight), the old Italian rules required:

$$L_{min} = 50\,m \tag{2.5}$$

2.2.2 Circular Curves

A rolling stock travelling a circular curve of radius R, at speed v is subject to the self-weight P and a horizontal force applied to the vehicle barycentre, named centrifugal force F_c (Fig. 2.3):

$$F_c = \frac{P}{g} \cdot \frac{v^2}{R} \tag{2.6}$$

The centrifugal force makes wheel *flanges* collide with the *gauge face of the* rails and, therefore, in certain conditions the rolling stock may risk derailment (see Chap. 4).

In curves very strong stresses insist on the track with consequent progressive deterioration of the original geometric configuration (misalignment) and wear out the inner side of the head of inner rail, mainly due to the slipping action of the wheels.

A further problem is linked to the travelling comfort perceived by passengers which can significantly reduce in curves as a consequence of the action of that force.

Fig. 2.3 Forces acting on a
train travelling on a
horizontal curve with a
superelevation (h)

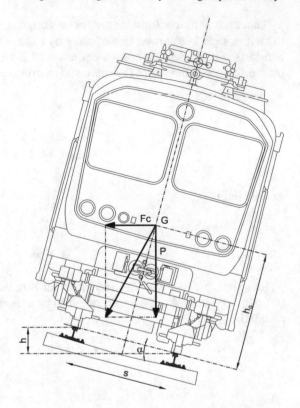

Keeping the previous inconveniences in mind, it is advisable to limit the intensity
of the centrifugal force and therefore to set, when designing, proper threshold values
of centrifugal acceleration a_c. To this purpose, the Italian RFI sets these limits [3]:

- $a_c = 0.6$ m/s^2 for heavy trains (freight trains and passenger trains composed of
  ordinary rolling stock);
- $a_c = 0.8$ m/s^2 for passenger trains composed of light material (electric locomotives) or wagons marked for $V \geq 140$ km/h, driven by locomotives E 646, E 636,
  E 632, E 633, E 444;
- $a_c = 1.0$ m/s^2 for passenger trains composed of rolling stock with high stability
  and low abrasiveness on tracks (e.g. E 444, ETR 250, ETR 300, ALE 601);
- $a_c = 1.8$ m/s^2 for tilting trains (e.g. the Italian *pendolino*).

Considering the centrifugal acceleration

$$a_c = \frac{v^2}{R} \tag{2.7}$$

the speed value V (in km/h) that, for a given radius R, leads to a prefixed value
acceleration, is obtained.

To provide a comfortable ride on a horizontal curve, generally the level of the outer rail is raised above the level of the inner rail. This is known as superelevation or cant [2] (cf. Fig. 3.2). In the absence of a superelevation on a curve, it then follows:

$$V = 3.6 \cdot \sqrt{a_c \cdot R} \quad [\text{km/h}] \tag{2.8}$$

In the presence of a superelevation (h) to compensate the components of the centrifugal and weight forces on the plane passing through the top surface of the rails (see Fig. 2.3), the following relation needs to be met:

$$P \cdot \text{sen}\,\alpha = F_c \cdot \cos\alpha; \quad \frac{F_c}{P} = \tan\alpha \tag{2.9}$$

Given that:

$$\frac{h}{s} = \text{sen}\,\alpha = \tan\alpha \cdot \cos\alpha \tag{2.10}$$

from the relations (2.9) and (2.10), it follows:

$$\frac{h}{s} \cdot \frac{\tan\alpha}{\text{sen}\,\alpha} = \frac{F_c}{P} \tag{2.11}$$

The value of the angle α is rather low, thus the ratio between its tangent and its sine is close to 1 (see Table 2.3).
Thus, it may be written:

$$\frac{h}{s} = \frac{F_c}{P} \tag{2.12}$$

Table 2.3 Characteristic values of the ratio tanα/sinα

h [mm]	s [mm]	h/s	A [rad]	α°	$\sin\alpha$	$\tan\alpha$	$\tan\alpha/\sin\alpha$
160	1500	0.10667	0.10687	6.12630	0.10667	0.10728	1.00574
140	1500	0.09333	0.09347	5.35812	0.09333	0.09374	1.00438
120	1500	0.08000	0.08009	4.59089	0.08000	0.08026	1.00322
100	1500	0.06667	0.06672	3.82449	0.06667	0.06682	1.00223
80	1500	0.05333	0.05336	3.05878	0.05333	0.05341	1.00143
60	1500	0.04000	0.04001	2.29361	0.04000	0.04003	1.00080
40	1500	0.02667	0.02667	1.52884	0.02667	0.02668	1.00036
20	1500	0.01333	0.01333	0.76435	0.01333	0.01333	1.00009
0	1500	0.00000	0.00000	0.00000	0.00000	0.00000	–

In which h is the superelevation, namely the difference in height between the outer and the inner rail on the curve. In Eq. (2.12) the s value is assumed to be equal to 1500 mm.

The previous Eq. (2.12) can also be written:

$$\frac{h}{s} = \frac{\frac{P}{g} \cdot \frac{v^2}{R}}{P} \tag{2.13}$$

From which the superelevation h can be deduced:

$$h = \frac{s}{g} \cdot \frac{v^2}{R} \tag{2.14}$$

By expressing h and s in mm, g in m/s², V in km/h and R in m, it yields:

$$h = 11.8 \cdot \frac{V^2}{R} \tag{2.15}$$

For reasons connected to the vehicle dynamics, in real situations h is limited to lower values than those inferred from Eq. (2.15); therefore by introducing the concept of non-compensated centrifugal acceleration a_{nc}, the previous Eq. (2.13) can be written as:

$$\frac{h}{s} = \frac{\frac{P}{g} \cdot \frac{v^2}{R} - \frac{P}{g} \cdot a_{nc}}{P} \tag{2.16}$$

which yields the h value corresponding to a preset threshold of non-compensated acceleration:

$$h = 11.8 \cdot \frac{V^2}{R} - \frac{s}{g} \cdot a_{nc} \tag{2.17}$$

The latter term of the previous expression is called the superelevation deficiency j; or:

$$j = \frac{s}{g} \cdot a_{nc} \tag{2.18}$$

In short, it follows:

$$h = 11.8 \cdot \frac{V^2}{R} - j \tag{2.19}$$

The Italian RFI sets values of non-compensated acceleration a_{nc} corresponding to the limits previously listed for a_c; therefore, by particularising them, it follows:

Table 2.4 Superelevation deficiency values

Non-compensated acceleration a_{nc} [m/s²]	Superelevation deficiency j [mm]
0.4	61
0.6	92
0.8	122
1.0	153
1.8	275

- $a_{nc} = 0.6$ m/s² for heavy trains (freight trains and passenger trains composed of ordinary rolling stock);
- $a_{nc} = 0.8$ m/s² for passenger trains composed of light material (electric locomotives) or wagons marked for $V \geq 140$ km/h, driven by locomotives E 646, E 636, E 632, E 633, E 444;
- $a_{nc} = 1.0$ m/s² for passenger trains composed of rolling stock with high stability and low abrasiveness on tracks (e.g. E 444, ETR 250, ETR 300, ALE 601);
- $a_{nc} = 1.8$ m/s² for tilting trains (e.g. the Italian *pendolino*).

The values of the superelevation deficiency j are shown in Table 2.4.

Considering that a given railway line is travelled by more or less fast trains, it can be associated to a certain heterotachic regime, corresponding to the interval between the minimum (V_1) and maximum (V_{max}) speeds of the so-called slow- and high-speed trains:

- V_{max} maximum speed of high-speed trains;
- V_1 minimum speed of slow-speed trains.

Apart from the superelevation deficiency j which can be associated to fast trains (travelling along the line at speed V_{max}), for the slow trains travelling with speed V_1 there is an excess in superelevation (e). In fact, for slow trains the weight force component which is parallel to the plane passing through the head of the rails (Fig. 2.3) prevails over the centrifugal force component in the same direction.

In short, in order to consider appropriate safety and comfort conditions for both fast and slow trains, the following Eqs. (2.20) and (2.21) must be contextually verified:

$$h = 11.8 \cdot \frac{V_{max}^2}{R} \tag{2.20}$$

$$h = 11.8 \cdot \frac{V_1^2}{R} + e \tag{2.21}$$

thus,

$$V_{max} = \sqrt{\frac{R}{11.8} \cdot (e + j) + V_1^2} \tag{2.22}$$

$$R = 11.8 \cdot \frac{V_{max}^2 - V_1^2}{e + j} \tag{2.23}$$

$$h = (e + j) \cdot \frac{V_{max}^2}{V_{max}^2 - V_1^2} - j \tag{2.24}$$

Given the scheduled traffic during the railway planning phase or, in other words, given V_{max} and V_1 values, j and e (or, alternatively, the non-compensated acceleration a_{nc} and the hyper-compensated acceleration a'_c), the curve R radius and the rail superelevation h can be calculated.

On this point, it is worth observing that when V_{max} increases, there is also an increase in construction costs, energy consumption and frequency of maintenance activities on lines.

When a_{nc} (or j) increases, the comfort of high-speed trains decreases; on the other hand, the radii of the circular curves and consequently both construction and maintenance costs rise.

In Italy RFI sets the maximum superelevation limit as $h_{max} = 160$ mm, which is prudential with respect to the value of $h = 180$ mm corresponding to the limit for a train casually stopped in curve to restart.

In ordinary lines the superelevation $h_{max} = 160$ mm is associated to the following values of minimum radius $R_{min}(V_{max})$:

$$V_{max} = 160 \, km/h \; a_{nc} = 0.60 \, m/s^2 \; R_{min} = 1.260 \, m$$
$$V_1 = 80 \, km/h \qquad a'_c = 0.65 \, m/s^2 \; h_{max} = 160 \, mm$$

On the other hand, for high-speed railway lines the following values are adopted:

$$V_{max} = 300 \, km/h \; a_{nc} = 0.60 \, m/s^2 \; R_{min} = 5450 \, m$$
$$V_1 = 80 \, km/h \qquad a'_c = 0.65 \, m/s^2 \; h_{max} = 105 \, mm$$

Since the horizontal alignment of a given railway line is nearly always formed from sections with curves of different radius, the value h_{max} is associated to R_{min}, while the superelevation is lower than the maximum one on curves of radii higher than R_{min}, namely:

$$If \, R > R_{min} \quad h < h_{max} \tag{2.25}$$

According to "Technical rules for designing railway lines" published by RFI in 2006 [4], the superelevation value h is obtained by imposing that the non-compensated acceleration is proportional to the superelevation itself. In other terms, setting $a_{nc} = 0.60$ m/s^2 for $h_{max} = 160$ mm, the acceleration value decreases when h decreases with the relation $a_{nc} \quad 0.60 \cdot (h/h_{max})$; therefore, it follows [3]:

$$h = \frac{160}{160 + 92} \cdot 11.8 \cdot \frac{V^2}{R} \tag{2.26}$$

From which:

- for ordinary railway lines:

$$h = 7.5 \cdot \frac{V^2}{R} \tag{2.27}$$

- for high-speed railway lines ($h_{max} = 105$ mm), similarly, it follows:

$$h = \frac{105}{105 + 92} \cdot 11.8 \cdot \frac{V^2}{R}; \quad h = 6.29 \cdot \frac{V^2}{R} \tag{2.28}$$

By means of Eqs. (2.27) and (2.28) the superelevation values h can be calculated when R and V vary (see Table 2.5 and Table 2.6) and R_{min} values can be estimated when V varies by imposing $h = h_{max} = 160$ mm in Eq. (2.27) and $h = h_{max} = 105$ mm in Eq. (2.28).

The abovementioned RFI rules [4] recommend adopting superelevation values lower than 160 mm in the following cases:

- on tracks adjacent to railway platforms, where the maximum superelevation must be 110 mm;
- in level crossings, bridges and tunnels where, in certain circumstances, reduced superelevation values can be used.

2.3 Speeds on Railway Lines

In railway engineering the term speed can be referred to numerous kinematic variables; but only the following are the most relevant for design purposes:

- *Limit speed* $\overline{V}$: is the speed used in travelling a circular curve of radius R with the superelevation hmax (160 mm) which determines a non-compensated centrifugal acceleration $a_{nc} = 0.6$ m/s². The value of this speed (see Table 2.7) is obtained from Eq. (2.17), considering that for 0.6 m/s², the superelevation deficiency $j = 92$ mm:

$$\overline{V} = 4.62 \cdot \sqrt{R} \tag{2.29}$$

Table 2.5 Superelevation values in function of radius and speed for ordinary railway lines

R [m]	Values of h [mm] for given V [km/h]										
	80	90	100	110	120	130	140	150	160	170	180
300	160	–	–	–	–	–	–	–	–	–	–
400	120	152	–	–	–	–	–	–	–	–	–
500	96	122	150	–	–	–	–	–	–	–	–
600	80	101	125	151	–	–	–	–	–	–	–
700	69	87	107	130	154	–	–	–	–	–	–
800	60	76	94	113	135	158	–	–	–	–	–
900	53	68	83	101	120	141	163	–	–	–	–
1000	48	61	75	91	108	127	147	–	–	–	–
1100	44	55	68	83	98	115	134	153	–	–	–
1200	40	51	63	76	90	106	123	141	160	–	–
1300	37	47	58	70	83	98	113	130	148	–	–
1400	34	43	54	65	77	91	105	121	137	155	–
1500	32	41	50	61	72	85	98	113	128	145	–
1600	30	38	47	57	68	79	92	105	120	135	152
1700	28	36	44	53	64	75	86	99	113	128	143
1800	27	34	42	50	60	70	82	94	107	120	135
1,900	25	32	39	48	57	67	77	89	101	114	128
000	24	30	38	45	54	63	74	84	96	108	122
2100	23	29	36	43	51	60	70	80	91	103	116
2200	22	28	34	41	49	58	67	77	87	99	110
2300	21	26	33	39	47	55	64	73	83	94	106
2400	20	25	31	38	45	53	61	70	80	90	101
2500	19	24	30	36	43	51	59	68	77	87	97
2600	18	23	29	35	42	49	57	65	74	83	93
2700	18	23	28	34	40	47	54	63	71	80	90
2800	17	22	27	32	39	45	53	60	69	77	87
2900	17	21	26	31	37	44	51	58	66	75	84
3000	16	20	25	30	36	42	49	56	64	72	81
3100	15	20	24	29	35	41	47	54	62	70	78
3200	15	19	23	28	34	40	46	53	60	68	76
3300	15	18	23	28	33	38	45	51	58	66	74
3400	14	18	22	27	32	37	43	50	56	64	71
3500	14	17	21	26	31	36	42	48	55	62	69
3600	13	17	21	25	30	35	41	47	53	60	68
3700	13	16	20	25	29	34	40	46	52	59	66

(continued)

Table 2.5 (continued)

R [m]	Values of h [mm] for given V [km/h]										
	80	90	100	110	120	130	140	150	160	170	180
3800	13	16	20	24	28	33	39	44	51	57	64
3900	12	16	19	23	28	33	38	43	49	56	62
4000	12	15	19	23	27	32	37	42	48	54	61
4100	12	15	18	22	26	31	36	41	47	53	59
4200	11	14	18	22	26	30	35	40	46	52	58
4300	11	14	17	21	25	29	34	39	45	50	57
4400	11	14	17	21	25	29	33	38	44	49	55
4500	11	14	17	20	24	28	33	38	43	48	54
4600	10	13	16	20	23	28	32	37	42	47	53
4700	10	13	16	19	23	27	31	36	41	46	52
4800	10	13	16	19	23	26	31	35	40	45	51
4900	10	12	15	19	22	26	30	34	39	44	50
5000	10	12	15	18	22	25	29	34	38	43	49

- *Route speed* V_t: considering a line section composed of varied circular curves, straight lines and transition curves, the route speed is defined as the speed associated to the curve with the smallest radius (R_{min}), characterised by superelevation hmax = 160 mm and a_{nc} = 0.6 m/s^2:

$$V_t = 4.62 \cdot \sqrt{R_{min}} \tag{2.30}$$

The curves with greater radius than the minimum one ($R > R_{min}$) must have superelevations inferior to the maximum value associated to the shortest curve (h < h_{max}); on them, therefore, a non-compensated centrifugal acceleration is generated with the following value:

$$a_{nc} = 0.60 \cdot \frac{h}{h_{max}} \tag{2.31}$$

Thus, the route speed clearly depends on the length of the line section under examination: for example, the Italian RFI considers the section lengths as greater or equal to 2000 m.

Along a railway line, the speed differential values associated to successive line sections conveniently need to be limited so as to provide users with a good travelling comfort.

For this reason, it is a good rule for every railway section marked by V_t = const. to have a minimum 2 km length. Moreover, between the consecutive sections i and

Table 2.6 Superelevation values in function of radius and speed for high-speed railway lines

R [m]	Values of h [mm] for given V [km/h]										
	200	210	220	230	240	250	260	270	280	290	300
2500	101	–	–	–	–	–	–	–	–	–	–
3000	84	92	101	–	–	–	–	–	–	–	–
3500	72	79	87	95	104	–	–	–	–	–	–
4000	63	69	76	83	91	98	–	–	–	–	–
4500	56	62	68	74	81	87	94	102	–	–	–
5000	50	55	61	67	72	79	85	92	99	106	–
5500	46	50	55	60	66	71	77	83	90	96	103
6000	42	46	51	55	60	66	71	76	82	88	94
6500	39	43	47	51	56	60	65	71	76	81	87
7000	36	40	43	48	52	56	61	66	70	76	81
7500	34	37	41	44	48	52	57	61	66	71	75
8000	31	35	38	42	45	49	53	57	62	66	71
8500	30	33	36	39	43	46	50	54	58	62	67
9000	28	31	34	37	40	44	47	51	55	59	63
9500	26	29	32	35	38	41	45	48	52	56	60
10,000	25	28	30	33	36	39	43	46	49	53	57
10,500	24	26	29	32	35	37	40	44	47	50	54
11,000	23	25	28	30	33	36	39	42	45	48	51
11,500	22	24	26	29	32	34	37	40	43	46	49
12,000	21	23	25	28	30	33	35	38	41	44	47
12,500	20	22	24	27	29	31	34	37	39	42	45
13,000	19	21	23	26	28	30	33	35	38	41	44
13,500	19	21	23	25	27	29	31	34	37	39	42
14,000	18	20	22	24	26	28	30	33	35	38	40
14,500	17	19	21	23	25	27	29	32	34	36	39
15,000	17	18	20	22	24	26	28	31	33	35	38
15,500	16	18	20	21	23	25	27	30	32	34	37
16,000	16	17	19	21	23	25	27	29	31	33	35
16,500	15	17	18	20	22	24	26	28	30	32	34
17,000	15	16	18	20	21	23	25	27	29	31	33
17,500	14	16	17	19	21	22	24	26	28	30	32
18,000	14	15	17	18	20	22	24	25	27	29	31
18,500	14	15	16	18	20	21	23	25	27	29	31
19,000	13	15	16	18	19	21	22	24	26	28	30
19,500	13	14	16	17	19	20	22	24	25	27	29

(continued)

Table 2.6 (continued)

R [m]	Values of h [mm] for given V [km/h]										
	200	210	220	230	240	250	260	270	280	290	300
20,000	13	14	15	17	18	20	21	23	25	26	28
20,500	12	14	15	16	18	19	21	22	24	26	28
21,000	12	13	14	16	17	19	20	22	23	25	27
21,500	12	13	14	15	17	18	20	21	23	25	26
22,000	11	13	14	15	16	18	19	21	22	24	26
22,500	11	12	14	15	16	17	19	20	22	24	25
23,000	11	12	13	14	16	17	18	20	21	23	25
23,500	11	12	13	14	15	17	18	20	21	23	24
24,000	10	12	13	14	15	16	18	19	21	22	24
24,500	10	11	12	14	15	16	17	19	20	22	23
25,000	10	11	12	13	14	16	17	18	20	21	23
25,500	10	11	12	13	14	15	17	18	19	21	22
26,000	10	11	12	13	14	15	16	18	19	20	22

Table 2.7 Limit speed values in function of the circular curve radii

R [m]	Limit speed [km/h]
150	56.58
300	80.02
450	98.00
600	113.17
750	126.52
900	138.60
1050	149.71
1200	160.04
1500	178.93
1800	196.01
2000	206.61

j with a rank speed (see next bullet point) equal to $V_{t,i}$ and $V_{t,j}$ respectively, the following condition must be satisfied: $\Delta V_{t(j-i)} = \left| V_{t,j} - V_{t,i} \right| \leq 60\,\text{km/h}$.

Table 2.8 Rolling stock types, rank and operational coefficients

a_{nc} [m/s²]	Rolling stock types	Rank	Operational coefficient
0.6	Freight wagons, old obsolete ordinary passenger coaches	A	4.62
0.8	Rolling stocks for passenger transport, for V ≥ 140 km/h	B	4.89
1.0	Passenger trains composed of anti-hunting oscillation systems	C	5.15
1.8	Tilting trains	P	6.07

- *Rank speed*: is the maximum speed by which a vehicle of a given rank[2] can travel a curve of minimum radius (R_{min}), being part of a line section of a given length. The speed is obtained in function of the non-compensated centrifugal acceleration value associated to the rank (and thus, to the characteristics of travelling train dynamics) and set by the rules below:

  – Rank A: non-compensated acceleration value $a_{nc} = 0.6$ m/s²;
  – Rank B: non-compensated acceleration value $a_{nc} = 0.8$ m/s²;
  – Rank C: non-compensated acceleration value $a_{nc} = 1.0$ m/s²;
  – Rank P: non-compensated acceleration value $a_{nc} = 1.8$ m/s².

For the above ranks the speed expressions can then be derived from Eq. (2.17):

$$V_A = 4.62 \cdot \sqrt{R_{min}} \tag{2.32}$$

$$V_B = 4.89 \cdot \sqrt{R_{min}} \tag{2.33}$$

$$V_C = 5.15 \cdot \sqrt{R_{min}} \tag{2.34}$$

$$V_p = 6.07 \cdot \sqrt{R_{min}} \tag{2.35}$$

The numerical coefficient in the previous four expressions is termed operational coefficient (see Table 2.8).

Relating the rank speed to the route speed, it follows[3]: $V_A = V_t$; $V_B = 1.06\ V_t$; $V_C = 1.11\ V_t$; $V_P = 1.31 V_t$.

---

[2] Rank A: freight wagons and old obsolete passenger coaches;

Rank B: rolling stocks for passenger transport still in use;

Rank C: rolling stocks equipped with anti-hunting oscillation systems (authorised to travel at a speed over 160 km/h), articulated electric train or electric unit train (ETR-type) with the exception of tilting trains;

Rank P: active tilting trains.

[3] Considering the values set for non-compensated acceleration, transverse jerk and rolling speed on transition curves, $V_A = 140$ km/h, $V_B = 160$ km/h, $V_C = 180$ km/h are set in RFI.

In short, a horizontal circular curve can be travelled with different maximum speeds. Therefore, the speed limits to be respected by vehicles of a given rank are shown by means of speed limit signs,[4] placed outside the railway track.

From Eqs. (2.32 and 2.35) the minimum radii of the circular curves (expressed in m) can be obtained when the rank varies:

$$R_{A\ min} = 0.0469 \cdot V_A^2 \tag{2.36}$$

$$R_{B\ min} = 0.0180 \cdot V_B^2 \tag{2.37}$$

$$R_{C\ min} = 0.0377 \cdot V_c^2 \tag{2.38}$$

$$R_{P\ min} = 0.0271 \cdot V_P^2 \tag{2.39}$$

- *Flank speed*: the maximum speed of a train depends on the parameters previously examined (non-compensated acceleration, circular curve radius, superelevation, etc.), on the other geometric alignment characteristics and restrictions required by rail traffic management systems, electric traction systems etc.

The term flank speed indicates the maximum permissible speed (set by the railway operating company) at which a train can travel along a line section of a given length; it may be lower or equal to the rank speed. By way of example, the railway track type can lead to a flank speed lower than the rank speed (e.g. 50 UNI rails require a Vc limit of 160 km/h).

- *Speed per hour:* is the train speed set in the train scheduling phase; it must be lower than the flank speed, so as to ensure adequate elasticity to the railway traffic flow.

The imposition of a lower speed than the flank speed enables it to build up a speed reserve ΔV ($\Delta V = Flank\ speed—Speed\ per\ hour$) which can be used to reduce train delays.

2.4 Transition Curves

In straight sections the two rails have the same height and, thus, the top surface of the rails (the rail plane) identifies a plane with a longitudinal gradient coinciding with the gradient of the axes of the two rails (or, in other words, with the longitudinal

---

[4] For a signpost with speed limits for ranks A, B and C (three panels), if rank C is not signalled (only two panels), the speed limit of rank C trains is the same as in rank B.

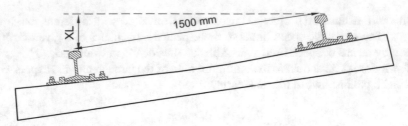

Fig. 2.4 Cross level

gradient of the axis of the track); therefore, if rails were ideally considered as two straight lines, they would turn out to be coplanar.

In transition curves, the inner and outer rails have a different height due to the superelevation; therefore, superelevation determines different values in the gradient of the two rails.

As a matter of fact, while the inside rail of the transition curve keeps the same height as in the track axis (as inferred from the longitudinal profile of the railway alignment), the outer rail is superelevated and consequently its longitudinal gradient is different from both the track axis and the inside rail. For this reason in transition curves, rails are ideally shaped as two skew lines in space, i.e. non-coplanar. The cross level XL is the difference in height between the top surfaces of the rails; it coincides with the height of the minor cathetus of the right triangle with a 1500 mm hypotenuse and a vertex angle equal to the angle between the running surface and a reference horizontal plane (see Fig. 2.4).

By the term twist γ [‰] is meant the inclination (expressed in per mille) referred to a rail towards the other, calculated as ratio between the difference of the cross level XL in two track sections at a given distance l, called measurement basis of the twist, and the distance itself. In accordance with the rules, the measurement basis can be 9 m or 3 m. In analytic terms:

$$\gamma = \frac{XL}{l} = \frac{h_1 - h_2}{l} \qquad (2.40)$$

Since along the transition curves the twist assumes values different from zero, while travelling, one of the four wheels of the bogie tends to unload or, in extreme cases, not to rest onto the rail, thus sometimes causing the vehicle derailment. Consequently, when railway lines are in operation, the twist limit can only be modified to the minimum extent: the limits for loaded tracks (measured with diagnostic trains or coaches, see Chap. 6) are shown in Table 2.9. If, after measurement, the twist overcomes the permissible limits, traffic restrictions must be applied in order to ensure safety.

Transition curves are inserted into horizontal railway alignment especially to improve passengers' travelling comfort and increase the railway infrastructure service life. These objectives are pursued by introducing:

Table 2.9 Twist limit values (loaded track measurements)

Twist limit values	
Measurement on 3 m basis	Measurement on 9 m basis
$\gamma_{3\,m} = 6.5\,\%o$	$\gamma_{9\,m} = 4.5\,\%o$

- on the horizontal plane a transition curve with a variable radius along its development. If the transition curve connects a straight line and a circular curve with radius R, its curvature radius holds values ranging between ∞ and R (increase in transverse acceleration during the motion on the transition curve). If, on the other hand, the geometric elements to be connected are two circular curves of radius R_1 and R_2, the curvature radius of the transition curve has values ranging along its length in the interval $[R_1; R_2]$;
- on the vertical plane a superelevation profile, along which the rail outer to the curve increases its height (compared to the inner rail), starting from the initial point of the horizontal transition curve until the desired vertical value (corresponding to the design superelevation h) at the initial point of the circular curve (Fig. 2.5).

In order to determine the length of the superelevation profile and, thus, the horizontal transition curve, it is necessary to define preliminarily three kinematic variables, more precisely:

- transverse jerk;

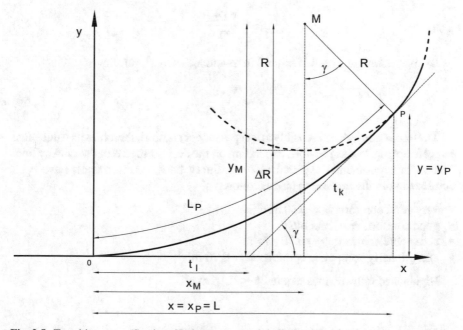

Fig. 2.5 Transition curve (Section 0P) between a straight line and the circular curve of radius R

- roll speed;
- lifting speed.

2.4.1 Transverse Jerk

The transverse jerk ψ [m/s^3] is the variation of non-compensated centrifugal acceleration in a time unit and represents the gradualness through which the non-compensated
centrifugal acceleration increases at a curve entrance, or decreases at a curve exit. The
transverse jerk is zero on straights—in that there is no centrifugal acceleration—and
also along the circular curves since the non-compensated centrifugal acceleration is
kept constant on these geometric elements. Analytically, it follows:

$$\psi = \frac{a_{nc}}{t} \tag{2.41}$$

Should, by hypothesis, the length l of the transition curve be travelled at a constant
speed v, then the travel time t will be:

$$t = \frac{l}{v} \tag{2.42}$$

From which it follows:

$$\psi = \frac{v \cdot a_{nc}}{l} \tag{2.43}$$

By expressing the speed in km/h, l in m and a_{nc} in m/s^2, it follows:

$$\psi = \frac{V \cdot a_{nc}}{3.6 \cdot l} \tag{2.44}$$

The transverse jerk can sensibly affect passenger comfort, therefore its value must
be preferably as low as possible, depending on the desired speeds of the railway line.
Thanks to experimental experiences on ordinary lines, comfort levels have been
correlated with the transverse jerk as follows:

- very good comfort: $\psi = 0.30$ m/s^3;
- good comfort: $\psi = 0.45$ m/s^3;
- acceptable comfort: $\psi = 0.70$ m/s^3;
- exceptionally acceptable comfort: $\psi = 0.70$ m/s^3;

High-speed railway lines require $\psi \leq 0.15$ m/s^3.

2.4.2 Roll Speed

Roll speed ω [rad/s] is the angular speed which the vehicle bogie reaches to change from the vertical position on the straight section to the tilted position in curve and vice versa. The maximum vehicle inclination α (Fig. 2.3) is obtained at time $t = l/v$, corresponding to the time necessary to travel the transition curve of length l at speed v. The following expressions must be taken into consideration (Fig. 2.3):

$$\alpha = \omega \cdot t \approx \frac{h}{s} = \frac{\omega \cdot l}{v} \tag{2.45}$$

$$tg\alpha \approx \frac{h}{s} = \frac{\omega \cdot l}{v} \tag{2.46}$$

Seeing that the longitudinal gradient p (evaluated on the vertical plan) of the transition curve of length l is:

$$p = \frac{h}{l} \tag{2.47}$$

it follows:

$$\omega = \frac{h \cdot v}{s \cdot l} = \frac{p \cdot v}{s} \tag{2.48}$$

By expressing speed in km/h, length l in m, h and s in mm, p in ‰, it follows:

$$\omega = \frac{h \cdot V}{3.6 \cdot s \cdot l} \tag{2.49}$$

$$\omega = \frac{p \cdot V}{3.6 \cdot s} \tag{2.50}$$

The roll speed must be properly limited in order not to enable the vehicle, once reached its final condition at the end of the transition curve, to keep on rotating by inertia as a consequence of a certain rotational kinetic energy acquired while travelling the transition curve.

2.4.3 The Lifting Speed

Alternative to the rolling speed, also the lifting speed V_s [mm/s] can be analysed: it corresponds to the vertical speed component of the outer wheel travelling along the superelevation profile in the transition curve. It is obtained from the roll speed ω (Eq. (2.49)) multiplied by the distance s (s = 1500 mm) between the contact points

Table 2.10 Limit values of the kinematic parameters for transition curves

a_{nc} [m/s²]	ψ [m/s³]	ω [rad/s]	V_s [mm/s]
0.6	0.25	0.036	54
0.8	0.35	0.038	57
1.0	0.40	0.040	60

of the two rails:

$$V_s = \frac{h \cdot V}{3.6 \cdot 1} \tag{2.51}$$

The lifting speed is limited to 25 –60 mm/s and exceptionally to 70 mm/s. In case of a linear altimetric connection (see Sect. 2.4.4) and a cubic parabola as transition curve, the following expressions is obtained:

$$\omega = \frac{h}{1500} \cdot \frac{\psi}{a_{nc}} \tag{2.52}$$

For this geometrical hypothesis, the limit values of transverse jerk, roll speed and lifting speed applied by the Italian RFI Rules are those shown in Table 2.10.

2.4.4 Superelevation Profile

Superelevation is run out or gained over the length of the transition curve (and not rather on the straight line or the circular curve).

In order to obtain the superelevation profile, the following design data are required:

(a) superelevation value h;
(b) transverse jerk value ψ;
(c) roll speed value ω;
(d) superelevation profile type.

The points (a), (b) and (c) have been widely dealt with in the previous sections; on the other hand, varied geometric configurations can be chosen for the superelevation profile (see Fig. 2.6):

- linear profile;
- biquadratic profile;
- sinusoidal profile.

The first has the advantage of being easily viable; the other two have the disadvantage of showing a higher maximum gradient but, on the other hand, they reduce the variations in both transverse jerk and roll speed at the entry and exit of the transition curve.

More often than not, the linear profile is also used for high-speed railway lines.

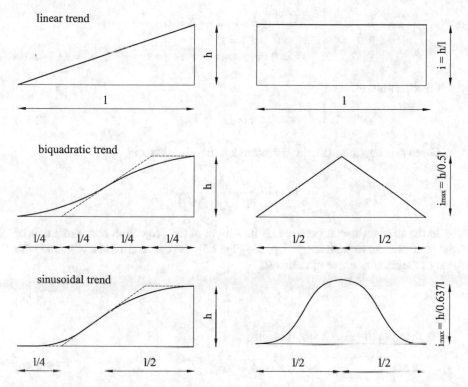

Fig. 2.6 Superelevation profile

2.4.5 Transition Curves: The Cubic Parabola

The cubic parabola is the traditional transition curve used in railways engineering. In a system of Cartesian axes (x, y) with the origin coinciding with the tangent point between the straight and the transition curve (see Fig. 2.5), the equation of the cubic parabola is as follows:

$$y = ax^3 + bx^2 + cx + d \tag{2.53}$$

where the first and second derivatives are given by the expressions:

$$y' = 3ax^2 + 2bx + c \tag{2.54}$$

$$y'' = 6ax + 2b \tag{2.55}$$

By imposing the boundary conditions:

$$y = 0 \quad \text{per } x = 0 \rightarrow d = 0$$
$$y' = 0 \quad \text{per } x = 0 \rightarrow c = 0$$
$$y'' = 0 \quad \text{per } x = 0 \rightarrow b = 0$$

this equation results:

$$y = ax^3 \tag{2.56}$$

Given the function y(x), the curvature k in the abscissa x is:

$$k(x) = \frac{y''(x)}{[1 + (y'(x))^2]^{\frac{3}{2}}} \tag{2.57}$$

In the abscissa $x = L$ (where L is the length of the projection onto the x axis of the parabolic curve with length Lp, see Fig. 2.5), the curvature must coincide with that of the circular curve of radius R:

$$k = \frac{1}{R} \tag{2.58}$$

By deriving the Eq. (2.56), it results:

$$y' = 3ax^2 \tag{2.59}$$

$$y'' = 6ax \tag{2.60}$$

Being $y'(x)^2 \ll 1$, Eq. (2.57) can be approximated with the relation $k = y''(x)$, thus:

$$y''_{(x=L)} = 6aL = k = \frac{1}{R} \tag{2.61}$$

$$a = \frac{1}{6RL} \tag{2.62}$$

In short, the equation of the cubic parabola is:

$$y = \frac{x^3}{6RL} \tag{2.63}$$

where, as previously said, L is the length of the projection of the parabolic curve onto the abscissa axis.

The tangent (tgγ) to the transition curve at a generic point of abscissa x (see Fig. 2.5) is obtained with the following expression:

$$tg\gamma = y' = \frac{x^2}{2RL} \tag{2.64}$$

The curvature ($k = 1/r$) is given by:

$$k = y'' = \frac{1}{r} = \frac{x}{RL} \tag{2.65}$$

It follows that, if travelled at a constant speed, the cubic parabola generates a centrifugal acceleration proportionate to the projection on the abscissae axis (x) of the corresponding curvilinear abscissa.[5]

The technical guidelines [4] provided by the Italian RFI introduce the parameter $A^2 = R \cdot L$. Thus, the previous expressions can also be written as:

$$y = \frac{x^3}{6A^2} \tag{2.66}$$

$$y' = tg\,\gamma = \frac{x^2}{2A^2} \tag{2.67}$$

$$y'' = \frac{1}{r} = \frac{x^2}{A^2} \tag{2.68}$$

The transition curves can be inserted in the horizontal alignment by applying the method of the "preserved centre" (with a reduction in the radius of the original circular curve) or the method of the "preserved radius" (where the radius of the circular curve does not change but the position of its centre does). The *preserved centre* method is the most frequently used but it implies a not always acceptable reduction (m) in the original radius R of the circular curve which needs connecting to the transition curve; such a reduction can be calculated with the relation:

$$m \approx \frac{L^2}{24 \cdot R} \tag{2.69}$$

Consequently, the radius of the definitive (deviated) circular curve R_f *is:*

$$R_f = R - m \tag{2.70}$$

The cubic parabola has different inconveniences especially due to the fact that its real length L_P usually is approximated by the length L of its projection onto the axis of the abscissas (cf. Fig. 2.5). Such an approximation involves some geometric discontinuities, better described in the following sections. For this reason, the clothoid is always suggested as a transition curve.

---

[5] Considering Eq. (2.65), the centrifugal acceleration can be expressed as: $a_c = \frac{v^2}{r} = \frac{v^2 \cdot x}{RL}$.

The length L of the parabola is determined differently in function of the railway line type:

- for lines with speeds lower than 160 km/h, the value of the gradient i (i = h/l) of the superelevation profile is evaluated in conformity with the underlying criteria:

  - i = 2.0‰ for V < 75 km/h;
  - i = 1.5‰ for V < 100 km/h;
  - i = 1.0‰ for V < 160 km/h.

From which the length L of the transition curve is obtained:

$$L = \frac{h}{i} \tag{2.71}$$

- for lines with speeds higher than 160 km/h the transition curve length is determined by imposing a transverse jerk as $\psi \leq 0.15$ m/s³. By Eq. (2.43), it follows:

$$L = \frac{v \cdot a_{nc}}{\psi} \tag{2.72}$$

2.4.6 Design of the Cubic Parabola with the Preserved Radius Method

The expressions explained in the previous section allow to obtain, as shown in [5], the length of the transition curve in function of the abscissa x and thus L_p in function of x_P (see Fig. 2.5):

$$L_p = x_P + \frac{1}{40} \frac{x_P^5}{A^4} - \frac{1}{1152} \frac{x_P^9}{A^8} \tag{2.73}$$

For small values of the angle γ, being $\tan\gamma \approx \gamma$, through expression (2.67), it results: $x = A \cdot \sqrt{2} \cdot \gamma^{\frac{1}{2}}$. From the latter and Eq. (2.73), it follows:

$$L_p = A \cdot \sqrt{2} \cdot (\gamma^{\frac{1}{2}} + \frac{1}{10} \cdot \gamma^{\frac{5}{2}} - \frac{1}{72} \cdot \gamma^{\frac{9}{2}}) \tag{2.74}$$

Moreover, the following expressions have to be considered:

$$x = A \cdot \sqrt{2} \cdot \tan\gamma \tag{2.75}$$

$$x_M = x_P - R \cdot \operatorname{sen} \gamma \tag{2.76}$$

$$y_M = y_P + R \cdot \cos \gamma \tag{2.77}$$

$$\Delta R = y_M - R \tag{2.78}$$

On the other hand, the long tangent t_l and the short tangent t_k (see Fig. 2.5) are calculated with the following relations:

$$t_l = x_P - \frac{y_P}{\operatorname{tg} \gamma} = L - \frac{L^3}{6RL} \cdot \frac{2RL}{L^2} = L - \frac{L}{3} = \frac{2}{3}L \tag{2.79}$$

$$t_k = \frac{y_P}{\operatorname{sen} \gamma} \approx \frac{y_P}{\operatorname{tg} \gamma} = \frac{L^3}{6RL} \cdot \frac{2RL}{L^2} = \frac{L}{3} \tag{2.80}$$

2.4.7 Design of the Cubic Parabola with the Preserved Centre Method

The curve length in function of the abscissa x or the angle γ is given from the relations (2.73) and (2.74); when γ varies, the abscissa x is calculated with the expression (2.75).

Remembering that $R_f = R - m$ (see expressions (2.69) and (2.70)), the value of the tangent γ_f at point P results to be different from the value of the original curve (of radius R), being

$$\operatorname{tg} \gamma_f = \frac{L^2}{2R_f L} = \frac{L}{2R_f} \tag{2.81}$$

Thus, after simple steps, it follows:

$$\operatorname{tg} \gamma_f = \frac{12RL}{24R^2 - L^2} \tag{2.82}$$

The coordinates $(x_M; y_M)$ of the centre of the circular curve are:

$$x_M = x_P - R_f \cdot \operatorname{sen} \gamma_f = L - R_f \cdot \operatorname{sen} \gamma_f \tag{2.83}$$

$$y_M = y_P + R_f \cdot \cos \gamma_f = \frac{L^3}{6A^2} + R_f \cdot \cos \gamma_f \tag{2.84}$$

It is worth noting that the values of the trigonometric tangent to the circular curve and to the cubic parabola at point P do not coincide (geometric discontinuity). In fact, the tangent at P to the circular curve is equal to [6]:

$$\text{tg } \alpha = \frac{L}{2 \cdot \sqrt{R_f^2 - \left(\frac{L}{2}\right)^2}} \tag{2.85}$$

Only at a first approximation, if we neglect $(L/2)^2$ with regard to R_f^2, the condition $\text{tg}\alpha = \text{tg}\gamma_f$ is met.

2.4.8   Transition Curves: The Clothoid

The clothoid equation is obtained by imposing that, along the length of the curve travelled at a constant speed, the transversal acceleration varies with the time linearly:

$$\frac{da_t}{dt} = \frac{d\left(\frac{v^2}{r}\right)}{dt} = \frac{v^2 d\left(\frac{1}{r}\right)}{dl} \cdot \frac{dl}{dt} = \frac{v^3 \cdot d\left(\frac{1}{r}\right)}{dl} = \text{const} = \psi \tag{2.86}$$

The constant ψ is the transverse jerk. From Eq. (2.86) it results:

$$d\left(\frac{1}{r}\right) = \frac{\psi}{v^3} \cdot dl \tag{2.87}$$

Thus,

$$r \cdot 1 = \frac{v^3}{\psi} = \cos t = A^2 \tag{2.88}$$

$$r \cdot 1 = A^2 \tag{2.89}$$

In the clothoid Eq. (2.89), r is the radius of the curvature at the point P of the curvilinear abscissa l (see Fig. 2.7) and A is the scale factor, also termed as clothoid parameter or scale parameter.

The clothoid is a special curve belonging to the spiral family of equations $r \cdot l^n = A^{n+1}$. The variation of n (form factor) generates the equations of some significant curves (n $= -1$ spiral; n $- 0$ circumference; n $= 1$ clothoid, n > 2 hyperclothoids; n $= \infty$ straight line).

By denoting with R the radius of the circular curve to be connected and with L the clothoid length,[6] when particularising Eq. (2.89) it results:

---

[6] It is worth recalling that for the cubic parabola L stands for the length of the projection onto the x axis whose length is L_p (see Fig. 2.5).

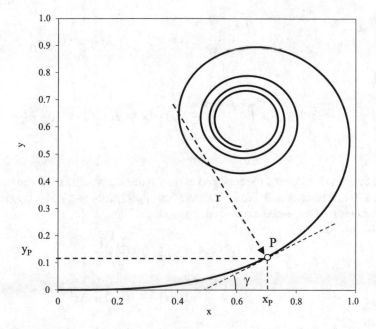

Fig. 2.7 The clothoid curve

$$R \cdot L = A^2 \tag{2.90}$$

The value of the tangent angle γ at a generic point of the curvilinear abscissa l is given by:

$$d\gamma = \frac{dl}{R} = \frac{1 \cdot dl}{A^2} \tag{2.91}$$

By integrating and considering that for $l = 0$, $\gamma = 0$, it follows:

$$\gamma = \frac{L^2}{2 \cdot A^2} = \frac{A^2}{2 \cdot R^2} = \frac{L}{2 \cdot R} \tag{2.92}$$

When γ varies, the coordinates can be obtained with the expressions $dx = \cos\gamma \cdot dl$ and $dy = \sin\gamma \cdot dl$. From Eq. (2.92) it also results:

$$dl = \frac{A \cdot \sqrt{2}}{2} \cdot \gamma^{-\frac{1}{2}} \cdot d\gamma \tag{2.93}$$

With the series expansion of the sine and cosine functions and their integration, the expressions[7] looked for x and y are obtained:

---

[7] By stopping at the first term of the series expansion of the functions (2.94) and (2.95), there follows: $x = A \cdot \sqrt{2\gamma}$; $y = A \cdot \sqrt{2\gamma} \cdot \frac{\gamma}{3} = x \cdot \frac{\gamma}{3}$. Considering Eq. (2.92) and approximating the

$$x = \int_0^\gamma \frac{A \cdot \sqrt{2}}{2} \cdot \gamma^{-\frac{1}{2}} \cdot (1 - \frac{g^2}{2!} + \frac{g^4}{4!} - \ldots) \cdot dg = A \cdot \sqrt{2 \cdot g} \cdot (1 - \frac{g^2}{10} + \frac{g^4}{216} - \ldots)$$

$$(2.94)$$

$$y = \int_0^\gamma \frac{A \cdot \sqrt{2}}{2} \cdot \gamma^{-\frac{1}{2}} \cdot (\gamma - \frac{\gamma^3}{3!} + \frac{\gamma^5}{5!} - \ldots) \cdot d\gamma = A \cdot \sqrt{2 \cdot \gamma} \cdot (\frac{\gamma}{3} - \frac{\gamma^3}{42} + \frac{\gamma^5}{1320} - \ldots$$

$$(2.95)$$

The coordinate values of the centre M of the osculating circumference in the point P (Fig. 2.7) can be obtained through the relations (2.92, 2.94 and 2.95). By stopping at the first term of the series expansion, it results:

$$x_M = A \cdot \sqrt{2 \cdot \gamma} - R \cdot \gamma \cong L - \frac{L}{2} = \frac{L}{2} \qquad (2.96)$$

$$y_M = A \cdot \sqrt{2 \cdot \gamma} \cdot \frac{\gamma}{3} + R \cdot (1 - \frac{\gamma^2}{2}) = R + \Delta r \qquad (2.97)$$

From Eq. (2.97), it follows:

$$\Delta r = A \cdot \sqrt{2 \cdot \gamma} \cdot \frac{\gamma}{3} - R \cdot \frac{\gamma^2}{2} = A \cdot \frac{A}{r} \cdot \frac{A^2}{6 \cdot R^2} - \frac{R}{2} \cdot \frac{A^4}{4R^4} = \frac{A^4}{24 \cdot R^3} = \frac{L^2}{24 \cdot R} \qquad (2.98)$$

For technical applications, the following relation is also highly interesting, in that it determines the parameter A of the clothoid for a given radius value of the circular curve R which is to be connected and for a given deviation ΔR between the straight and a circular curve:

$$A = \sqrt[4]{24 \cdot R^3 \cdot \Delta R \cdot (1 + \frac{3}{14} \cdot \frac{\Delta R}{R})} \qquad (2.99)$$

The long tangent t_l and the short tangent t_k (see Fig. 2.5) are obtained from expressions (2.79) and (2.80). The previous relations are applicable to both the preserved radius and the preserved centre methods.

The length of the clothoid is determined with the same criteria previously described for the cubic parabola and here reproduced for the reader's convenience:

- for lines with speeds lower than 160 km/h, the value of the gradient i (i = h/l, see Fig. 2.6) of the superelevation junction is set and assumed equal to:

  - i = 2.0‰ for V < 75 km/h;
  - i = 1.5‰ for V < 100 km/h;

---

length L of the curve to its projection L_x onto the abscissa axis, it follows $y = \frac{x^3}{6 \cdot R L_x}$ which is the equation of the cubic parabola.

Table 2.11 Minimum permissible length of straights and circular curves

Railway line type	V ≤ 200 km/h	200 ≤ V ≤ 300 km/h
Limit values [m]	$\frac{V_{max}}{3}$ (never below 30 m)	$\frac{V_{max}}{1.5}$
Exceptional values [m]	$\frac{V_{max}}{5}$ (never below 30 m)	$\frac{V_{max}}{2.5}$

- $i = 1.0‰$ for $V < 160$ km/h.

From which the length L of the transition curve is obtained:

$$L = \frac{h}{p} \tag{2.100}$$

- for lines with speeds higher than 160 km/h, the transition curve length is determined by imposing a transverse jerk as $\psi \leq 0.15$ m/s^3:

$$L = \frac{v \cdot a_{nc}}{\psi} \tag{2.101}$$

2.5 Minimum Permissible Length of Straights and Circular Curves

The Italian guidelines [4] set limit values for the minimum lengths of straight and circular curve sections as shown in Table 2.11, differentiated according to the line speeds (lower or higher than 200 km/h). No indication is given for the length of transition curves.

If for technical reasons the minimum length of a straight section cannot be ensured (see Table 2.11), then it has to be eliminated in order to connect directly the two circular curves to the transition curves without the straight section, as represented by the schemes in Fig. 2.8.

2.6 The Vertical Alignment

The projection of the railway alignment onto a vertical plane is a line composed of a series of straight sections with constant-gradient (inclination along the railway

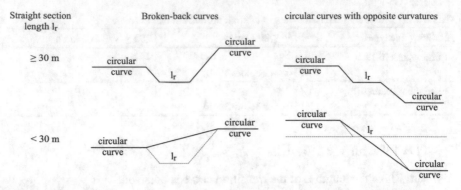

Fig. 2.8 Diagram of the curvature (k = 1/r) for different values of the straight section length (l_r) [3]

longitudinal axis), known as grades, properly connected by circular curves or, more rarely, by parabolic curves (altimetric or vertical curves).

2.6.1 Maximum Gradient and Minimum Length of Grades

The gradients of grades are attributed with the following rule:

- positive gradient: if the heights of the vertical alignment increase when the chainages increase;
- negative gradient: if the heights of the vertical alignment decrease when the chainages increase;

The suggested absolute values of maximum gradient are:

- 5 –8‰ for railway lines on level terrain;
- 15 –18‰ for railway lines on rolling terrain;
- 20 –25‰ for railway lines on mountainous terrain;
- 30 –35‰ for secondary railway lines specialised for passenger trains;
- 10‰ for tunnels;
- 1‰ for stations (preferably 0‰).

For secondary railway lines in the mountains, it is sometimes advisable to design grades with a gradient higher than 35‰; rack systems are required for travelling on such steep sections (see Chap. 15).

Of great interest for design purposes is also the limit or braking gradient, defined as the value of the longitudinal gradient downwards and on a straight section along which the vehicle (initially at a standstill) starts moving forward if it is not braked.

Since the specific rolling resistance r_l (see Chap. 1) for modest speeds assumes the value of 2 –3 daN/t for some vehicle types, the limit gradient il is exactly equal to 2 –3‰; in fact, higher gradients generate a tractive force greater than the rolling

Table 2.12 Maximum gradient values

Railway line or section types	$i_{max}(\%_o)$
Lines for only passenger trains	35
Mixed traffic lines	12
Terminals	1.2
Stations	10

resistance (see Chap. 1). Such a value is therefore the maximum theoretical gradient to attribute to the tracks at stations. However, in order to ensure a further safety margin, it is advisable not to overcome the 1‰ or, even better, 0‰ gradient.

The Italian rules [4] provide the maximum gradient values (i_{max}) shown in Table 2.12. Only in exceptional cases higher values can be used, after permission by the railway operator.

In order to choose the grades on alignments with horizontal curves of radius below 1000 m, the effect of the additional curve resistance, as well as the grade resistance, to the train motion should be taken into consideration (cf. Chap. 1).

The length of every grade L_g, [m] net of the length of the vertical curve, must be higher than the following minimum value (L_{min}) proportional to the speed V:

$$L_g \geq L_{min} = \frac{V}{1.8} \tag{2.102}$$

with a binding minimum value of 30 m.

2.6.2 Vertical Curves

Vertical curves are conventionally named as follows:

- convex curves, if they have convexities downwards;
- concave curves, if they have concavities upwards.

The geometric conditions identifying the vertical curve type are summarised in Table 2.13, where i_j denotes the gradient of the grade j and i_{j+1} denotes the gradient of the grade $j + 1$ (see Fig. 2.9), according to the rules of the sign attributed to gradients (see previous section).

Table 2.13 Type of vertical cruves

Gradient of the grades	Vertical curve
$i_{j+1} < i_j$	Convex
$i_{j+1} > i_j$	Concave
$i_j > 0, i_{j+1} < 0$	Crest (special type of convex curves)
$i_j < 0, i_{j+1} > 0$	Sag (special type of concave curves)

Fig. 2.9 Gradients and
vertical curve

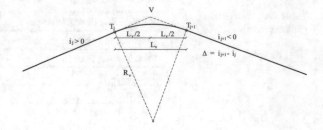

With reference to Table 2.13, it is worth recalling that sag curves are not recommended along cutting cross sections, in that rainwater drainage would be extremely hard.

The vertical curves are usually made with circumference arches (see Fig. 2.9), without any transition curves.

In order to ensure travelling comfort, the Italian RFI sets a vertical acceleration limit value av [4]. It is conventionally assumed that:

$$a_v \leq 0.5 \ \left[m/s^2\right] \tag{2.103}$$

Therefore, the value of the radius of the vertical curve must be:

$$R_v \geq \frac{V^2}{a_v} \tag{2.104}$$

By expressing the maximum line speed in km/h, a_v in m/s^2 and R_v in m, it results:

$$R_v = \frac{V_{max}^2}{12.96 \cdot a_v} \geq (R_v)_{lim} \tag{2.105}$$

For a long time in Italy the following minimum values have been used:

- $R_V = 3000$ m for ordinary railway lines;
- $R_V = 25,000$ m for high-speed railway lines.

The 2006 guidelines [4] set the $(R_v)_{lim}$ values which can be calculated with the expressions given in Table 2.14.

By denoting with $\Delta i = |i_{j+1} - i_j|$ the gradient difference between the two grades to connect (see Fig. 2.9), the projection of the vertical curve length onto a horizontal plane is:

$$L_v = R_v \cdot \Delta i \tag{2.106}$$

The condition $L_v \geq 20$ m must be respected if the difference between the gradients is:

- $\Delta i > 2‰$ for speeds up to 230 km/h;

Table 2.14 Prescriptive limits on radius values $(R_v)_{lim}$ for vertical curves

Lines	Up to 200 km/h	$200 < V \leq 300$ km/h
Limit value suggested [m]	$0.35\ V^2_{max}$	$0.35\ V^2_{max}$
Minimum limit value [m]	$0.25\ V^2_{max}$ [b]	$0.175\ V^2_{max}$ [a]

[a] with a + 10% tolerance on convex curves and + 30% tolerance on concave curves

[b] never below radii of 2000 m

- $\Delta i > 1\%_o$ for speeds over 230 km/h.

2.7 Three-Dimensional Alignment

Horizontal and vertical alignment coordination is the design phase devoted to selecting the mutual correspondence between geometric elements of the horizontal sections (straights, circular curves, transition curves) and vertical sections (grades and vertical curves). For ordinary railways the following are prohibited [4, 7]:

- overlapping of transition curves with altimetric curves, otherwise vehicle instability events and rail torsion stresses may occur;
- grades changes in presence of horizontal curves and track devices such as switches and crossings (Chap. 7).

2.8 Double-Track Railway Line

The distance between the longitudinal axes of the two tracks in double-track railway lines (see Figs. 3.2 and 3.3, Chap. 3) must be greater or equal to the minimum values shown in Table 2.15.

Table 2.15 Minimum distance between the longitudinal axes of the two tracks in double-track railway sections

Type of railways	Minimum distance [m]
Railway line modernization with max speed ≤ 200 km/h	3555
New lines with max speed ≤ 200 km/h	4000
New lines with max speed ≤ 250 km/h	4200
New lines with max speed between 250–300 km/h	4500

Finally, it is worth noting that the railway alignment in terminals, stations, cross-roads, etc. (see Chap. 8) must respect specific technical guidelines [8], necessary to refer to for any detailed study.

References

1. Esveld C (2001) In: Modern railway track. 2nd edn. MRT-Productions
2. Chandra S, Agarwal MM (2007) In: Railway engineering, Oxford University Press
3. Bono G, Focacci C, Lanni S (1997) Railway track (in Italian, *La sovrastruttura ferroviaria*), CIFI
4. *Technical rules for designing railway lines* (code: RFI TCAR IT AR 01 001 A). Italian RFI (in Italian)
5. Kellogg NB (1899) The transition curve or curve of adjustment. Press of Upton Bros, Printers and publishers
6. Agostinacchio M, Ciampa D, Olita S (2005) Roads, railways and airports (in Italian, *Strade ferrovie aeroporti*), EPC libri
7. Policicchio F (2007) Railway infrastructures (in Italian. Firenze University Press, Lineamenti di infrastrutture Ferroviarie)
8. Design of new railway alignments (2013) (code: RFI TCAR IT AR 01 003 A), Italian RFI (in Italian)

Chapter 3
The Railway Track

Abstract This chapter deals with the tenets involved in the construction of ballasted track (conventional track) and ballastless track (slab track). In addition, it provides a synthetic description of railway track components as well as materials, characteristics and properties.

The term "railway track" or "railway superstructure" entails tracks, switches, crossings and ballast beds. More specifically, the railway track is the superstructure placed on the top of the railway body. The traditional railway track basically consists of a flat framework—obtained by properly linking rails, sleepers and fasteners—and a layer of crushed stone called ballast. Below the latter, if necessary, a further layer called sub-ballast is placed (e.g. a bituminous conglomerate layer). On the other hand, slab tracks are commonly used on high-speed railway lines and light rail systems.

Thus, railway tracks are usually subdivided into the following types (Fig. 3.1):

- traditional, with ballast (ballasted track, see cross sections in Figs. 3.2, 3.3 and 3.4;
- innovative or unconventional (ballastless track or slab track), in which the traditional elastic combination of sleepers and ballast is generally replaced by a rigid layer of concrete.

The ballasted tracks have rather low construction and maintenance costs. However, under the action of traffic loads they tend to lose their original track geometry and deteriorate ballast aggregates, thus reducing the friction coefficient and the capacity of damping vibrations and noise generated by train passages.

The ballastless or slab tracks are composed of rails, fasteners and a reinforced concrete slab or precompressed reinforced concrete slab which is placed on a concrete foundation. Compared to the ballasted ones, the slab tracks have much higher construction costs but lower maintenance costs. Their further advantage is the reduction in height and weight and, if rails are elastically supported by rail pads, also in noise and vibrations. Several factors influence the choice for the proper railway track type such as train axial loads, traffic intensity, travelling speeds, desired useful life, etc.

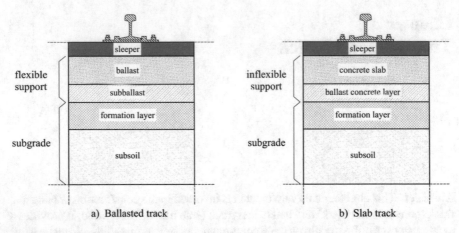

a) Ballasted track   b) Slab track

Fig. 3.1 Ballasted and ballastless (slab) tracks

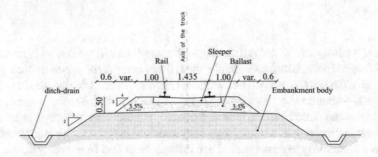

Fig. 3.2 Section type for a simple track line and speeds below 160 km/h (ballast volume 1.27 m³/m)

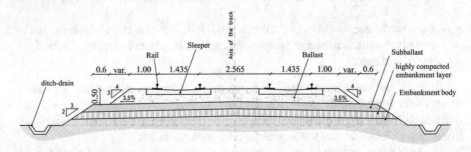

Fig. 3.3 Section type for a double track line and speeds below 160 km/h (ballast volume 3.93 m³/m)

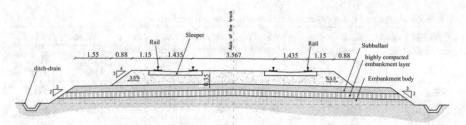

Fig. 3.4 Section type for high speed lines (ballast volume 5.57 m^3/m)

3.1 Rails

Rails are made of structural high-carbon steel in order to support and guide the train wheels. On ordinary and high-speed railways, the flat-footed rail, also called Vignoles profile, is the most frequently used rail type (see Fig. 3.5), but other rail profiles are also commonly employed, e.g. double-headed rails, grooved rails (used in urban tramways), switch rails, Burback rails, etc.

Rails are tilted inward at an angle of 1 in 20 to reduce wear and tear on the rails as well as on the tread of the wheels. Due to its characteristic inverted T-type cross section, the Vignoles profile can be fixed directly to the sleepers with the help of spikes.

Table 3.1 illustrates sizes and main mechanical characteristics of some frequently used rails. The UIC 60 rail is generally used in case of high-traffic load tracks (daily traffic load > 35,000 t); on the other hand, the UIC 50 is employed for low-traffic load tracks (daily traffic load < 25,000 t). However, the Vignoles rail profile, which is the most widespread nowadays and also employed for high-speed railway networks, is the UIC 60.

The relation between the weight per unit length M (expressed in kg/m) and the section area A (expressed in mm^2) is:

$$M = 0.00786 \cdot A \tag{3.1}$$

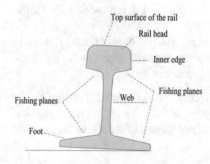

Fig. 3.5 Vignoles rail profile

Table 3.1 Sizes and main characteristics of Vignoles rail profiles

UNI EN 13,674 denomination	Old Denomination	Weight [kg/m]	Height [mm]	Foot width [mm]	Area [cm²]	Moment of inertia	
						J_x [cm⁴]	J_y [cm⁴]
46E4	FS 46	46.9	145	135	59.78	1688	338.6
50E5	UIC 50	49.86	148	135	63.62	1844	362.4
60E1	UIC 60	60.21	172	150	76.70	3038.3	512.3

Table 3.2 Steel standards provided for by the Italian RFI

Steel grades	HBW hardness range	Description
R260	260–300	Non-heat-treated non alloy Carbon- Manganese (C-Mn)
R320Cr	320–360	Non-heat-treated alloy (1% Cr)
R350HT	350–390	Heat-treated non alloy Carbon-Manganese (C-Mn)

Rails are made from hot rolled steel in the following main profiles: 12, 24, 36, 48 and 108 m. The profiles are then jointed (jointed rails) or welded in situ (continuous welded rails, CWR) to one another.

Theoretically, the longer the rail, the lesser the number of joints and fastenings required and the lesser the cost of railway track construction and maintenance. But if, on the one hand, longer rails provide easy and comfortable rides, on the other, their length is shortened to cope with, for instance, heavy internal thermal stresses or difficulties in transport facilities and in getting greater expansion joints.

The group of structural steels can belong to classes 700, 900A (the most used class), 900B, 1100, 1200 and 1200 HH [1]. The class number value denotes yield strength expressed in N/mm².

For different markets steel rails can be classified into various standards, such as GB, UIC, DIN, ASTM, AREMA.

The most recent technical specifications for railway material and equipment supply at the Italian RFI, published in 2014 [1], detail the steel properties to be used for rail construction as shown in Table 3.2.

Steel mechanical properties and chemical composition are provided for in the European Standard UNI EN 13,674-1.

As previously stated, rails are placed with a vertical axis tilted inward at an angle of 1 in 20, in order to enable them to match properly to coach wheels with the same inclination. The wheel coning and the rail inclination ensure:

- the reduction of the self-excited vibration mechanism, known as "hunting";
- the optimum behaviour along horizontal curves: in fact, the trajectories travelled by internal and external wheels of each wheelset have different lengths and this

---

[1] Italian specifications for rails and switches "RFI TCAR SF AR 02 001 C".

Table 3.3 Rail type, tilt and gauge values

Railway	Rail type	Tilt	Gauge and tolerances [mm]
Network rail	BS113A	1:20	1432^{-0+3}
DB AG	UIC60	1:40	1435^{-0+3}
FS	UIC60	1:20	1435^{-2+1}
NS	UIC54, NP46	1:40/1:20	1435^{-1+3}
NSB	S49, UIC54	1:20	1435^{-3+3}
OBB	S49, UIC54, UIC60	1:40	1435^{-2+2}
SBB	UIC60	1:40	1435^{-2+2}
SJ	SJ43, SJ50	1:30	1435^{-3+3}
SNCF	UIC60	1:20	1436^{-2+2}

means that, along curves, each wheel rolls on a rolling rim of a different diameter (see Chap. 2).

Although the 1:20 tilt is the most used worldwide, some railway companies require different tilting values (see Table 3.3 [1]).

3.1.1 Rail Joints and Welding of Rails

Rails are made of steel and consequently very susceptible to temperature changes which cause longitudinal expansions or contractions, partially impeded by fastening systems and sleepers. The direct fastenings make it necessary to introduce an allowable clearance (expansion gap) between two subsequent rail profiles to be connected longitudinally, with the aim of ensuring the expected expansions and contractions during the railway track life cycle.

The maximum clearance between the rail ends allowed by the Italian RFI is 14 mm (intended to be provided during initial assembly); therefore, the rail length between successive joints is traditionally limited to 12 m (the free expansion of a rail for a temperature variation of 70° is of the order of 10 mm). Generally, indirect K-type fastening systems enable the use of longer rail profiles.

If rail connection systems (e.g. welded rails) or available clearances should not allow for the required expansion, a non-negligible stress state is produced in the bar which can sometimes cause the rail break (for tensile stresses) or misalignment (for compressive stresses).

In railway construction techniques "short jointed rails" (SJRs) are usually distinguished from continuous welded rails (CWRs). In a SJR variations in rail length are permitted by joints; on the other hand, in CWRs their lack brings about inevitable tensile stresses which should be examined properly in the design phase of the railway. Thus, in order to avoid breaks and severe geometrical faults, the environmental

Fig. 3.6 A suspended joint

temperature differences should be estimated between the welding process and the new railway line operation. The railway track construction technique depends on economic evaluations (i.e. costs regarding materials, equipment, maintenance and depreciation) and on desired comfort levels. In SJRs (Fig. 3.5) rail ends are connected with a joint made of two fishplates, screws, nuts and bolts.

Joints can be of two very different types, supported or suspended. In supported joints, the joint has a sleeper directly under the rail ends. In suspended joints (Fig. 3.6) the joint is between two sleepers and the rail is partly cantilevered at the joint.

Although the supported joint is preferable to the suspended one, in that it ensures a better continuity of the rolling surface, a break in rail continuity also occurs at the joint, thus affecting travelling comfort, especially at high speeds. This is the reason why the continuous welded rails are more and more frequently used, in that their rolling surfaces are devoid of discontinuities. Rail welding can take place in a plant or in situ with flash-butt or aluminothermic welding process. Flash-butt welding technique is applied by connecting the end faces of the rails being joined, which have been previously brought to plastic state by means of intense electric currents.

In the aluminothermic welding process molten steel is poured into a mould surrounding the gap between the rail ends to be joined. The principle of aluminothermic welding is based on an exothermic chemical reaction of aluminium powder and iron oxide, producing sufficient heat to cause melting. The filler metal is melted in a crucible placed on the rail ends to be welded. After cooling the welded area, the crucible, moulds and extra material consisting especially of aluminium oxide with a glass-like structure (corundum) are removed. When the welding process is completed, the necessary quality controls are applied with ultrasonic equipment.

Between the two techniques, the flash-butt welding guarantees more on time resistance and is also the fastest, taking only three minutes for each welding against twenty minutes required for aluminothermic welding.

3.2 Sleepers

Sleepers are structures positioned transversally to the track axis which support rails. The latter are connected to sleepers with fastenings. Rails, sleepers and fastenings form the flat framework of the railway track.

Sleepers can be made of wood (oak, beech, chestnut, cerris or pine, 60 to 100 kg in weight and 30-to-40-year service life [1]), concrete (50 years' service life [1]) or, more rarely, in steel, rubber or rubber-coated concrete.

In addition to ensuring the gauge, sleepers have further functions:

- to anchor the flat framework to the ballast, resist mechanical stresses (i.e. accelerations and decelerations) from rolling stocks and thermal stresses (especially in continuous welded rails);
- to distribute the rolling stock loads on the ballast.

The sleeper spacing is generally set in 60 cm, which is a value representing a good compromise between opposite requirements in terms of railway track performance as well as construction and maintenance costs. In lines with highly intense traffic or heavy freight traffic, sometimes shorter sleeper spacing (up to 50 cm) can be adopted. The Italian RFI used to apply 66-cm moduli to old branch lines. Today, regardless of the line type, it uses a constant 60-cm sleeper spacing with a plus or minus 3-cm tolerance, which is necessary, for instance, to adjust the sleeper position to joints. The classification of wooden sleepers is defined by the UIC Technical Specifications and is based on cross section sizes with only three profiles. In any case, sleepers must not be wider than 30 cm and longer than 2.60 m.

As said above, the function of sleepers is to distribute loads from the rails to the ballast. The pressure distribution on the ballast sections depends upon the size and shape of the ballast and the degree of consolidation. In particular, the pressure distribution on the ballast under the sleepers may be estimated with the simplified model[2] schematised in Fig. 3.7 and with the following relations:

$$
\begin{aligned}
l' &= z + 2.7 \cdot a \quad \text{(for new sleepers)} \\
l' &= z + 1.5 \cdot a \quad \text{(for worn - out sleepers)}
\end{aligned}
\tag{3.2}
$$

in which:

- l' is the width over which the load is distributed at depth a;
- z is the sleeper width;
- a is the depth of the layer under the sleeper soffit of which the pressure is to be estimated.

According to their type, prestressed and non-prestressed concrete sleepers vary in weight ranging from 200 to 380 kg.

---

[2] Although the lines of equal pressure are bulb-shaped, yet for simplicity, the load dispersion can be assumed to be linear with a given inclination to the vertical.

Fig. 3.7 Simplified pressure
distribution on the ballast

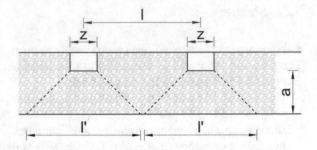

They provide a flat framework with high stiffness and increase the resistance to transversal stresses and service life of the track. For these reasons they are generally applied to new lines but also to the oldest ones which are subject to railway renewal. From a functional viewpoint, mono-block and two-block sleepers are very different, as explained in the following sections.

3.2.1 *Mono-block Prestressed Concrete Sleepers*

A mono-block sleeper is manufactured by only one concrete block, prestressed longitudinally. They are distinguished by the compression method into post-tensioned sleeper and pre-tensioned sleeper. The sizes of mono-block sleepers vary from country to country. For instance, the American Railway Engineering Association sets precise limits: length 2.74 m, base width 33 cm, height 25.4 cm, weight 360 kg. For several years the Italian RFI has by now installed pre-compressed reinforced concrete sleepers with trapezoidal cross sections, 2.30 m long and weighing 250 –280 kg, in railway lines with speeds lower than 200 km/h (see Fig. 3.8). Vibrating mechanical means are then necessary to consolidate a ballasted truck with concrete sleepers. Nowadays numerous types of mono-block sleepers are available. For instance, German railways (Deutsche Bahn AG) frequently use profiles B 70 W, B 90 and B 75 [1].

3.2.2 *Twin-Block Reinforced Concrete Sleepers*

A twin block sleeper (Fig. 3.9) consists of two reinforced concrete blocks joined through a steel rod or rigid steel beam. Compared to mono-block sleepers, a twin-block sleeper gives greater resistance to lateral forces but, on the other hand, it tends to bend under loads, thus slightly increasing the gauge. It is commonly used in France, Spain, Belgium, Portugal, Greece, Mexico, Brazil, Tunisia and Algeria. For instance, in France the twin-block sleeper B450 (260 kg in weight) is used on TGV railway lines (speeds up to 300 –350 km/h).

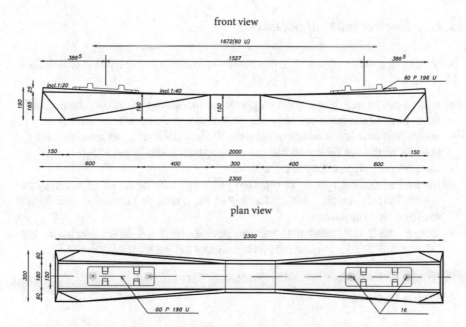

Fig. 3.8 Reinforced concrete mono-block sleeper (Italian RFI type)

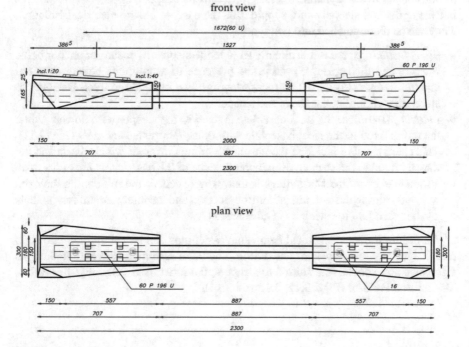

Fig. 3.9 A twin-block sleeper

3.2.3 Unconventional Sleepers

According to specific railway line characteristics, other sleeper types may be of help [2 –5], e.g.:

*   *wide sleepers* and *frame sleepers*, useful for increasing the resistance of the flat framework of the track and reducing pressures and deformations on ballast;
*   *mono- and two-block concrete sleepers* EGA and HP-BB, able to improve the stability of the flat framework of the track against lateral forces [5];
*   steel sleepers (types: Sw9 and Sw7DRB);
*   sleepers in natural rubber, in recycled polypropylene or recycled polyethylene (types: TieTek, Axion, I-Plas, KLP, MPW, etc.), able to increase durability and resistance to corrosion;
*   sleepers with integrated piezoelectric devices installed below the rails (type "*Green Rail*"), able to generate electric energy during train travelling.

3.3 Fastenings

Fastenings are devices that allow rails to be anchored to sleepers. Their function is to transmit static and dynamic stresses from rails to sleepers, to prevent movements between rails and sleepers and, should it be the case, to ensure electrical isolation. They can be distinguished into two types:

*   *direct fastenings*: the rail is directly fixed to the sleeper by means of anchor bolts, washers and dish springs, even in the presence of a baseplate between the rail and the sleeper (Fig. 3.10). Direct fastenings also include the track fastening to structures without ballast beds and sleepers;
*   *indirect fastenings:* the steel baseplate is fixed to the sleeper with dowels, while the rail is fixed to the steel baseplate with bolts, fish plates and nuts (Fig. 3.11). This category also includes the elastic fastenings through which the rail is not directly fixed to the sleeper, even in the absence of the baseplate. One of the main advantages of indirect fastenings is that, when needed, the rail can be removed without undoing the fastenings from the sleepers and the intermediate components can be placed on the sleepers in advance [2].

According to the mechanical behaviour, fastenings can be subdivided into rigid (e.g. K type fastening system) and elastic. The latter fastenings are able to dampen vibrations and shocks transmitted to sleepers; the most used fastenings are Nabla, Vossloh and Pandrol (Figs. 3.12, 3.13 and 3.14).

Fig. 3.10 Direct fastening

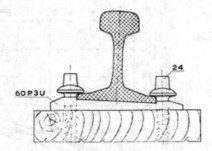

Fig. 3.11 Indirect fastening

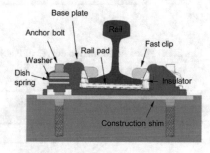

Fig. 3.12 Nabla

Fig. 3.13 Vossloh

Fig. 3.14 Pandrol

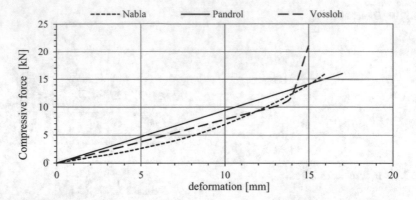

Fig. 3.15 Force–deformation curves of Nabla, Vossloh and Pandrol fastenings [6]

Table 3.4 Clamping load of fastenings K, Nabla, Vossloh and Pandrol [6]	Fastening	Clamping load [kN]
	K	10
	Nabla	11
	Vossloh	10.5
	Pandrol	13

These fastenings are fixed with the help of springs which control the deformation behaviour, partly influenced by the rubber pads below the rails (if present). The Pandrol clip (also known as elastic rail clip) can be installed mechanically with specific equipment (clipping machine), able to fasten and unfasten the clip with fast execution speed. The force–deformation curves of the elastic fastenings are illustrated in Fig. 3.15, on the other hand Table 3.4 shows the typical clamping loads of rigid and elastic fastenings.

Among the new-generation elastic fastenings there are, for example, Pandrol Fast Clip, Pandrol Safelock, Pandrol Vipa and the Pandrol Vanguard System (Fig. 3.16). The latter supports the rail by the rail web rather than from underneath the rail foot. This means that remarkably low vertical stiffness (less than 4 kN/mm) can be achieved, without compromising the control of lateral rail head movement and dynamic track gauge. The device reduces noise emission when used on bridge and viaduct structures thanks to the high level of vibration isolation between the rail and the bridge/viaduct below. It can be applied to metros also in curves with a small radius up to 40 m.

Fig. 3.16 Pandrol Vanguard fastening

3.4 Rail Pads

The rail pads are laid between rails and sleepers and made up of elastic material. They are aimed at transferring the rail load to the sleeper while filtering out the high frequency force components with advantages in terms of vibrations and noise reduction. Thus, they must offer damping properties (i.e. elevated hysteresis loop surface area) but low deformation under load.

A rail pad normally has a service life between 5 and 10 years. The most frequently used thicknesses are 9.5 –10 mm.

In conformity with the EN 134,812, rail pads can be subdivided into:

- soft: degrees of stiffness $\leq$ 80 kN/mm;
- average: 80 kN/mm < degrees of stiffness $\leq$ 150 kN/mm;
- hard:degrees of stiffness > 150 kN/mm.

3.5 Ballast Bed

The ballast bed is a layer of broken stones, gravel, or any other granular material with a given thickness placed and packed below and around sleepers for distributing the load from the sleepers to the formation. As a result of internal friction between the grains, the ballast can absorb considerable compressive stresses, but not tensile stresses. Ballast beds (see Figs. 3.2, 3.3, 3.4) aim to:

- distribute vertical loads transmitted from sleepers to a large area of the formation;
- provide the resistance to the track for longitudinal and lateral stability;
- keep the track alignment unchanged during the railway service life;
- absorb the vertical and horizontal stresses borne by the track during coach travelling;
- give elasticity and resilience to the track;
- provide effective drainage to the track.

The compacted bulk density of ballast material shall not be less than 1.5 t/m³ and the internal friction angle must not be less than 45°. Aggregates must have a fist shape with sharp edges.

The most used rocks are basalt, porphyry and granite (service life of 20–30 years). On the other hand, river gravel and rounded aggregates cannot be used. The original roughness and roundness of granular material must be maintained over time. Traditional Italian railway lines require a Los Angeles index lower than 20–25; for high-speed railway lines, such an index must be lower or equal to 16. Railway operators generally prescribe granular materials ranging between 15–20 and 60–65 mm in size.

Table 3.5 shows particle-size distributions required in the UK, Germany, India and Australia [7, 8]; Fig. 3.17 illustrates the particle-size distribution limits used by the Italian RFI.

Table 3.5 Comparison between particle-size distributions in various countries

UK		Germany		India		Australia	
Size	Cumulative	Size	Cumulative	Size	Cumulative	Size	Cumulative
[mm]	% Passing	[mm]	% Passing	[mm]	% Passing	[mm]	% Passing
63	100	63	100	65	95–100	63	100
50	70–100	50	65–100	40	40–60	53	85–100
40	30–65	40	30–65	20	0–2	37.5	20–65
31.5	0–25	31.5	0–25	–	–	26.5	0–20
22.4	0–3	25	–	–	–	19	0–5
32–50	≥ 50	–	–	–	–	13.2	0–2
–	–	–	–	–	–	4.75	0–1
–	–	–	–	–	–	0.075	0–1

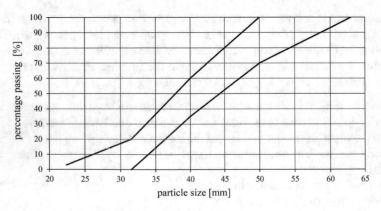

Fig. 3.17 The Italian RFI particle size distribution limits

In order to avoid water stagnation, the formation (substructure) below the ballast must have a falling camber from the axis of the track to the two edges, with a transversal gradient of 3.5%.

The thickness of the ballast cushion depends on the axle load, sleeper type, sleeper spacing and other related factors; the standard thickness, measured below the sleepers, is 35 cm.

In some countries, the thickness "a" of the ballast cushion is calculated in function of the sleeper spacing "l" and each sleeper's width "z". For instance, the Indian railways [9] use the following expression:

$$a = \frac{1 - z}{2} \qquad\qquad (3.3)$$

With a minimum thickness of 15 –20 cm.

In Italy the minimum thickness of the ballast cushion, evaluated below the sleepers, is 25 cm (20 cm for tunnels). Ballast properties must comply with the European Standard UNI EN 13,450 set out in 2013.

3.6 The Sub-ballast and the Over Compacted Subgrade Soils

In order to limit the railway track deterioration, the structure of a typical track foundation is formed with a sub-ballast layer placed over compacted subgrade soils. The over compacted layer (placed under the ballast and the sub-ballast), with a minimum 30 cm thickness, is built on the top part of the formation.

The sub-ballast is used in order to distribute the load evenly over the formation. It consists of a layer normally made of granular materials with a percentage of fine sand (0.06 –0.2 mm) higher or equal to 20% (technique especially used in Germany, the UK and Switzerland) or of two layers, the higher with a fine sand percentage of 30 –80% and the lower with the gravel (technique used in France) [6]. Generally, the sub-ballast should never be less than 150 mm thick.

In Italy and in other countries, e.g. in Japan, for high-speed railways (HSR) a cement-treated granular layer or bituminous-treated granular layer (see Table 3.6) is used as a sub-ballast. For HSR the sub-ballast is 12 cm thick if consisting of bituminous-treated granular material or 20 cm thick if it is made up of cement-treated granular material and laid on an over compacted subgrade layer with a minimum thickness of 30 cm.

Some railways do not use a sub-ballast layer but rather a larger thickness of the formation layer above the subgrade. In the absence of the sub-ballast, geotextile membranes are usually introduced below the ballast, thus working as a separator between the subgrade and the track and preventing ballast aggregates from filtering through the subgrade and, vice versa, the finest soil particles from moving upwards into the ballast. From a constructive viewpoint, it is good practice to build a sand layer

Table 3.6 Characteristics of
the two sub-ballast types [6]

Item	Bituminous-treated granular layer	Cement-treated granular layer
Grain sizes	0–20 mm	0–40 mm
Bitumen/cement content	4%	3%
Thickness	12 cm	20 cm
Resistance in traction	23 bar	10.4 bar
Elastic modulus E	40,000 bar at 30 °C, 90,000 bar at 20 °C	3000 bar

Fig. 3.18 Elastic sub-ballast
mat for bridges and tunnels

of at least 5 cm thickness on geotextile membranes and avoid them to be punched
by the ballast.

3.7   The Elastic Sub-ballast Mat

In order to provide the railway track with greater elasticity, on bridges and tunnels one
or more elastic mats[3] can be used on condition that they have an adequate thickness
(e.g. 25 mm if in natural rubber and 45 mm if in polyurethane elastomers [6]). The
mats soften vibrations and noises and reduce the stresses transmitted to the structure
(Fig. 3.18).

---

[3] Anti-vibrating mats are characterised with a variable density ranging between 75 –170 kg/m^3,
resistance to compression between 14 and 50 kPa, static elasticity modulus between 0.15 and
0.85 N/mm^2 and dynamic elasticity modulus between 0.35 N/mm^2 and 3.6 N/mm^2.

3.8 Subgrade and Formation

The formation of the railway infrastructure may be in an embankment or a cutting depending on the rail level and general configuration of the area.

3.8.1 Embankment Sections

Embankments are required when the railway track height is greater than the natural (existing) soil level. In railway embankments the elements listed below are identified:

- laying surface: generally, a 30-cm-thick blanket is provided at the top of the formation;
- anti-capillary layer: built on the formation with such characteristics to prevent water from rising by capillarity;
- body: a 30-cm-deep series of overlapped layers and compacted soil layers which allow the railway track to rest on, at a height superior to that of the land level;
- last embankment layer or over-compacted layer: the layer lying below the ballast or the sub-ballast, if it is present.

The Italian Ministerial Decree 14 September 2005 on "Technical Standards for Construction", whose details of interest will be mentioned below, establishes that considering the poor tolerances permitted by the railway track, in order to ensure rolling stocks to travel safely, the embankment must have only limited deformations.

Railway embankments must be constructed so that the residual subsidences are not 10% higher than theoretical subsidences calculated during the railway design process and, in any case, below 5 cm. As shown in the railway cross sections of Figs. 3.2, 3.3 and 3.4, the embankment slope is generally set at two-thirds and is covered with a 30 cm vegetal topsoil layer to allow grassing over. Very high embankments (H>6.00 m) require berms with a minimum width of 2.00 m for every 6.00 m embankment height.

The embankment body is formed of soils from excavations, foundation or tunnel excavations belonging to groups A1, A2-4, A2-5, A2-6, A2-7, A3 and A4, as provided for in AASHTO and CNR-UNI 10,006 regulations (see Table 3.7), as well as from borrow pits belonging to the same groups. There must not be soils from group A3 with a uniformity coefficient lower than 7 [10, 11]. Soils must be laid down, one layer on top of another; their thickness must not be higher than 50 cm for soil groups A1 and A2-4 and more than 30 cm for soil groups A2-5, A2-6, A2-7, A3 and A4 (see Table 3.7). Each layer must be well compacted to reach a dry density equal to 95% of Modified AASHTO Density (i.e. laboratory density); afterwards, another overlapping layer can be laid over it. Each layer must have a transversal 3.5% inclination. In each layer of the embankment body, the value of the bearing capacity evaluated by the deformation modulus must result to be, on the basis of a plate load test according to CNR-BU N. 146, higher than 20 MPa for embankment areas with a below 1 m distance from the lateral edges and more than 40 MPa for the remaining central area.

3.8.2 Cut Sections

The technical modalities to excavate trenches are described in several railway regulations, for instance in [10]. The in situ soil can be used as laying of the over compacted layer only if belonging to groups A1, A3 (with a uniformity coefficient higher than 7) or A2-4 under the AASHTO classification (see Table 3.7). It must be compacted so as to reach a dry density not lower than 95% of the maximum density, obtained for that soil through the modified AASHTO compaction test. The deformation modulus, measured at the first load cycle must not be lower than 40 MPa and the ratio between the moduli of the 1st and 2nd cycles must not be lower than 0.60. If this is not the case, a proper deep-thick blanket must be excavated and the filling up soil must ensure the minimum module value of 20 MPa for all the related layers, except for the upper layer, i.e. that one forming the supporting surface of the over compacted soil whose minimum modulus value, measured at the first load cycle, must not be lower than 40 MPa.

3.9 The Slab Track

In this superstructure type the flat framework of the track is built on a concrete slab, either prefabricated or made on site (see Fig. 3.19). There follow small deformations, deteriorations and reduction in maintenance costs compared to ballasted tracks, but with higher initial construction costs. Also, its service life is 50 –60 years, against 30 –40 years of ballasted tracks [1].

The following slab track systems are among the most well-known (Fig. 3.20) [2, 6]:

- *the Japanese system*: rails are directly clamped to a prefabricated reinforced concrete or precompressed reinforced concrete platform with specific fastenings placed every 62.5 cm. There are two slab types: one for cold environment (made of precompressed reinforced concrete, sized 493 × 234 × 19 cm), the other for mild environment (made of reinforced concrete, sized 495 × 234 × 16 cm or 19 cm). The platform rests on a rubber mat, 5 mm thick, with a 5 cm cement mortar layer below it. Its concrete foundation rests on a mechanically stabilized soil. Slabs are bound to each other by cylindrical stoppers;
- *the Italian system:* the prefabricated precompressed reinforced concrete slab is sized at 475 × 250 × 16 cm. Below it there are i) an anti-vibrant resilient mat,

Table 3.7 Soil classification according to AASHTO and CNR UNI 10,006 regulations

General classification	Granular materials (35% or less passing nr. 200 sieve (0.075 mm)							Silt–clay materials More than 35% or less passing nr. 200 sieve (0.075 mm)			
Group classification	A-1		A-3	A-2				A-4	A-5	A-6	A-7
	A-1-a	A-1-b		A-2-4	A-2-5	A-2-6	A-2-7				A-7-5 A-7-6
(a) Sieve analysis: percent passing											
(i) 2.00 mm (n. 10)	50 max	...									
(ii) 0.425 (n. 40)	30 max	50 max	51 min	...	...	...	...	...	...	...	...
(iii) 0.075 (n. 200)	15 max	25 max	10 max	35 max	35 max	35 max	35 max	36 min	36 min	36 min	36 min
(b) Characteristics of fraction passing 0.425 mm (nr. 40)											
(i) Liquid limit	...		...	40 max	41 min	40 max	41 min	40 max	41 min	40 max	41 min
(ii) Plasticity index	6 max		N.P	10 max	10 max	11 min	11 min	10 max	10 max	11 min	11 min[a]
(c) Usual types of significant constituent materials	Stone fragments, gravel or sand		Fine sand	Silty or clayey gravel and sand				Silty soil		Clayey soil	
(d) General rating as subgrade	Excellent to good							Fair to poor			

[a] If plasticity index is equal or less than (Liquid limit—30), the soil is A-7-5 (i.e. PL > 30%)
If plasticity index is greater than (Liquid limit—30), the soil is A-7-6 (i.e. PL < 30%)

Fig. 3.19 A slab track

25 mm thick, previously glued under the platform and ii) a bituminous mortar of 5 cm. A reinforced concrete foundation, 25 cm thick, is implemented in embankment and cut sections. At the two ends of the slab there are two semi-cylindrical stoppers (50 cm diameter and 20 cm height) fitted into the foundation. The weight of this railway track is markedly inferior to that of the ballasted track (1150 and 5300 kg/m respectively for double track lines, with 4 m interaxle), thus bringing clear advantages in reducing the permanent load when used in bridges and viaducts;

- *the PACT (Paved Concrete Track) system*: its reinforced concrete platform is laid down on a foundation or ballast resting on a 15 m thick concrete layer. Some neoprene plates are fitted between the rail and the slab (10 mm) with Pandrol e-Clip fastenings;
- *the STEDEF system:* characterised for the use of two-block sleepers, 12 m elastomer plates, Nabla fasteners and precompressed reinforced concrete or reinforced concrete platform;
- *the RHEDA system:* characterised for the use of mono-block sleepers B 70, embedded in a 42 cm reinforced concrete baseplate laid down and Vossloh Ioarv 300 fastenings. In open sky sections (i.e. embankment and cut sections), a 30 cm-high lean concrete foundation is implemented below the baseplate, reinforced longitudinally and transversally. In tunnel sections there is no foundation for the presence of the reverse arch and the reinforced concrete baseplate is 32 cm in height;
- *the RHEDA system 2000*: different from the previous one only for the twin-block sleepers (B 365 W60M) and for the constructive technique applied to the platform.

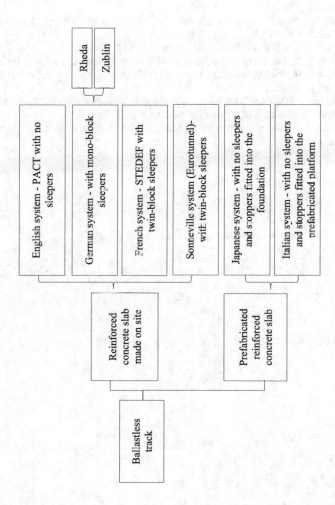

Fig. 3.20 Ballastless track types

References

1. Lichtberger B (2010) In: Track compendium. Eurail Press
2. Esveld C (2001) In: Modern railway track, 2nd edn. MRT-Productions
3. Ferdous W et al (2015) Composite railway sleepers—recent developments, challenges and future prospects. Compos Struct 134:158–168
4. Guerrieri M, Ticali D (2012) Ballasted track superstructures: performance of innovative railway sleepers. Civil-Comp Proc 98
5. Nazmul H (2017) Threshold radius of a ballasted CWR curved track: curve classification. J Transp Eng Part A: Syst 143(7)
6. Bono G, Focacci C, Lanni S (1997) In: Railway track. CIFI (in Italian)
7. Guler H, Mert N (2012) A Comparative Analysis of Railway Ballast. Civil-Comp Proceed 98
8. Aggregates and Rock For Engineering Purposes (1996) Part 7: Railway ballast. Standards Australia, NSW, Australia
9. Chandra S, Agarwal, MM (2007) In: Railway engineering, Oxford University Press
10. Italia RFI. Capitolato costruzioni opere civili. Section V (in Italian)
11. Santagata FA et al (2016) Road construction (in Italian, *Strade: teoria e tecnica delle costruzioni stradali*). Pearson

Chapter 4
Wheel-Rail Interaction and Derailment Analysis

Abstract This chapter deals with analysis the contact area between wheel and rail and the pressure distribution obtained by applying Hertz's theory. Since railway accidents result in heavy loss of life and property, the derailment risk levels according to Nadal's formula are also analysed in this chapter.

4.1 The Contact Area Between Wheel and Rail

The wheel-rail interaction plays a pivotal role in analysing microscopic and macroscopic phenomena such as adhesion, derailment dynamics, track deterioration, generation and propagation of noise and vibrations.

Hertz's theory allows to analyse the pressure distribution and deformation statuses taking place when two curved elastic bodies, in this case wheel and rail, are compressed one against the other. This theory can be applied only if the following conditions are met:

- elastic, homogeneous and isotropic materials;
- absent or negligible friction forces;
- small-sized contact area between the two bodies;
- interaction forces orthogonal to the contact area.

Hertz's theory shows that the contact surface between wheel and rail has an elliptical form (see Fig. 4.1) and the stress distribution over the contact surface has a semi-elliptical form.

By denoting the main curvature radii of body 1 (wheel) with R_1, R_1', and the modulus of elasticity and Poisson's ratio of the body 1 material respectively with E_1 and v_1, and by indicating the corresponding geometrical and mechanical variables of body 2 (rail) with R_2, R_2', E_2 and v_2, and the compression force (vertical wheel load) with Q, the contact ellipse semi-axes a and b (with a > b) as projection of the contact area wheel-rail onto the plane tangential to the two surfaces (see Fig. 4.1) can be obtained from the following expressions [1]:

$$a = 0.908 \cdot \mu(\theta) \cdot \sqrt[3]{\frac{Q}{D} \cdot \frac{1}{E^*}} \tag{4.1}$$

© The Author(s), under exclusive license to Springer Nature Switzerland AG 2023
M. Guerrieri, *Fundamentals of Railway Design*, Springer Tracts in Civil Engineering,
https://doi.org/10.1007/978-3-031-24030-0_4

Fig. 4.1 Contact area
between wheel and rail on
the plane tangential to the
two surfaces and pressure
distribution

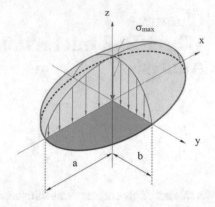

$$b = 0.908 \cdot \eta(\theta) \cdot \sqrt[3]{\frac{Q}{D} \cdot \frac{1}{E^*}} \tag{4.2}$$

Since

$$\frac{1}{E^*} = \frac{1-v_1^2}{E_1} + \frac{1-v_2^2}{E_2} \tag{4.3}$$

$$D = \frac{1}{2} \cdot \left(\frac{1}{R_1} \pm \frac{1}{R_2} + \frac{1}{R_1'} \pm \frac{1}{R_2'} \right) \tag{4.4}$$

In Eq. (4.4) the curvatures are to be considered positive if the curvature centre is inside the body (convex surface) and negative in the opposite case (concave surface).

By denoting with ϕ the angle between the planes α and β containing the two maximum (or minimum) curvatures of bodies 1 and 2 respectively [1], with

$$C = \frac{1}{2} \sqrt{ \left(\frac{1}{R_1} - \frac{1}{R_1'} \right)^2 + \left(\frac{1}{R_2} - \frac{1}{R_2'} \right)^2 + 2 \cdot \left(\frac{1}{R_1} - \frac{1}{R_1'} \right) \cdot \left(\frac{1}{R_2} - \frac{1}{R_2'} \right) \cdot \cos(2\varphi) } \tag{4.5}$$

and with

$$\theta = \arccos \frac{C}{D} \tag{4.6}$$

$\mu(\theta)$ and $\eta(\theta)$ values can be assessed with the diagram in Fig. 4.2, and consequently the semi-axes a and b can be calculated with the expressions (4.1) and (4.2).

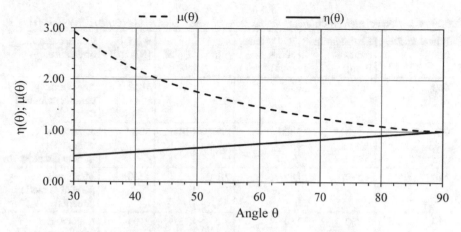

Fig. 4.2   Values of coefficients $\mu(\theta)$ and $\mu(\theta)$

The mean pressure σ_m and the maximum pressure σ_{max} acting both on the contact area (see Fig. 4.1) are equal to.
[1][1, 2]:

$$\sigma_m = \frac{Q}{\pi ab} \tag{4.7}$$

$$\sigma_{max} = \frac{3}{2}\sigma_m = \frac{3}{2}\frac{Q}{\pi ab} \tag{4.8}$$

The maximum stress point lies below the contact surface at a depth of $(0.2–0.5)b$. Table 4.1 shows the values of the ellipse semi-axes and compressive stresses for different combinations of wheel and rail curvatures and for a vertical wheel load Q = 60 kN [2].

(a)   (b)   (c)

Fig. 4.3   Contact area for wheel ϕ 500 mm and rail JIS 50 (**a**), wheel ϕ 260 mm and rail UNI 60 (**b**), wheel ϕ 960 mm and rail JIS 50 (**c**)

---

[1] The pressure distribution on the contact area surface has a semi-elliptical form. In the reference system of Fig. 4.1, it results: $\sigma(x, y) = \sigma_{max}\sqrt{1-\left(\frac{x}{a}\right)^2 + \left(\frac{y}{b}\right)^2}$.

Table 4.1 Contact area for different combinations of wheel and rail curvatures (Q = 60 kN) [2]

Wheel Radius [mm]	Flange profile radius [mm]	Rail profile radius [mm]	a [mm]	b [mm]	σ_{max} [N/mm^2]	Contact area orientation
460	∞	300	6.1	4.7	1012	Major axis parallel to motion direction
460	−330	300	3.9	14.6	502	Major axis orthogonal to motion direction
460	−330	80	7.1	2.7	1520	Major axis parallel to motion direction
150	−330	80	4.2	3.3	2103	Major axis parallel to motion direction

The modern pulse-echo ultrasonic techniques confirm the results from Hertz's theory as to contact area shape and size, with the only exception of the cases in which wheel and/or rail surfaces present irregularities [3].

4.2 The Adhesion Modifiers

The characteristic values of the wheel-rail adhesive coefficient have been examined in Chap. 1. Tables 4.2 and 4.3 list, synthetically, the main advantages and disadvantages associated to adhesive coefficients of elevated and low values.

Table 4.2 Main advantages and disadvantages related to high adhesive coefficient values

Elevated values of the adhesive coefficient	
Advantages	Disadvantages
Reduced noise	Elevated energy consumption
Small braking distance	Higher derailment risk
Very high tractive force	−

Table 4.3 Main advantages and disadvantages related to low adhesive coefficient values

Low values of the adhesive coefficient	
Advantages	Disadvantages
Reduced energy consumption	More frequent deteriorations
Lower derailment risk	Elevated noise
Major wheel/rail service life	Difficulty in overcoming grades With elevated gradient

While in the past the technique for increasing the adhesive coefficient in specific alignment sections (e.g. grades with steep slopes) was to sprinkle sand on rails, nowadays some products directly act on the longitudinal and transverse sliding of the wheel, thus influencing the adhesion coefficient indirectly. They are generally polymeric films which significantly reduce the sliding and, thus, increase the adhesive coefficient (up to values of the order of 0.35, see Chap. 1); they also prolong the service life of the rail by slowing the development of surface corrugations especially along the horizontal curves with low radius [4, 5].

Other circumstances ask for the adhesive coefficient reduction, for instance to lower the derailment coefficient below proper safety thresholds (see § 4.3) by employing, in this case, the so-called LCF (low coefficient of friction) lubricant products. They are devices attached to the bogie frame that, being equipped with a solid stick in contact with the wheel flange, spread a fine film over rims. The additive is spread when the rims reach high temperatures, that is, at high speeds.

4.3 The Derailment Risk and Nadal's Formula

The derailment is one of the most common railway accidents and is often due to a series of concomitant causes (e.g. worn rails, damaged wheels, high travelling speed, poor track geometry etc.). Its dynamics can nearly always be traced back to one of the following elementary cases:

- wheel climb on the rail: the wheel flange tends to overcome the rail as a consequence of the high transversal forces during the wheel/rail contact;
- wheel slide-up (slide-up derailment): the wheel falls off from the rail along the motion direction; for instance, in case of an irregular gauge increase;
- rail jump-up: the wheel jumps over the rail after an extremely violent and sudden collision between wheel and rail.

The wheel climb derailment, also called climbing derailment, is undoubtedly the most frequent and occurs in three distinct phases (see Fig. 4.4):

- phase A: the wheel flange is in contact with the vertical surface of the head of the rail;
- phase B: the wheel flange starts climbing onto the surface of the head of the rail;
- phase C: the wheel flange moves on the surface of the head of the rail. Should this be the case, the wheel is no more driven along the rail and its trajectory deviates from the alignment of the track.

In order to analyse the derailment dynamics, the wheel-rail contact is assumed to be not widespread on an elliptical area (see Figs. 4.1 and 4.3) but of a point type.

At the instant of derailment, when the wheel flange is in the phase of climbing onto the head of the rail, certain forces act on the contact point between rail and wheel. The interaction force F (i.e. the normal reaction of the rail) between the two bodies (see Fig. 4.5) can be subdivided into the two components N (normal force)

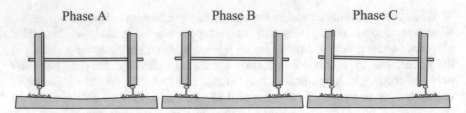

Phase A   Phase B   Phase C

Fig. 4.4 Elementary phases of the wheel climb derailment

Fig. 4.5 Forces at the wheel/rail contact point

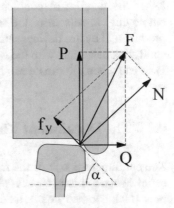

and f_y (tangential force), the former orthogonal, the latter parallel to the tangential plane in the wheel-rail contact point, inclined to the horizontal plane at an angle α (flange angle). F can be also subdivided into vertical direction P (reaction of the rail to the instantaneous wheel load) and horizontal direction Q (flange force). P and Q are estimated through a load analysis (see Chap. 5).

The ratio Q/P is known as the derailment coefficient. In a limit equilibrium condition, it results:

$$\frac{Q}{P} = \frac{\tan \alpha - \frac{f_y}{N}}{1 + \frac{f_y}{N} \cdot \tan \alpha} \tag{4.9}$$

The adhesion is guaranteed if $f_y \leq f \cdot N$. The limit value is $f_y = f \cdot N$. From Eq. (4.9) it yields that, in limit adhesion conditions, the derailment coefficient can be calculated with the expression:

$$\frac{Q}{P} = \frac{\tan \alpha - f}{1 + f \cdot \tan \alpha} \tag{4.10}$$

The coefficient of friction f between the wheel flange and the rail cannot be confused with the adhesive coefficient dealt with in Chap. 1.

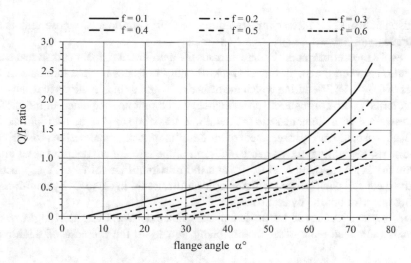

Fig. 4.6 Derailment coefficient according to Nadal's equation

Equation (4.10) is the so-called Nadal's equation and provides a crucial criterion for assessing the rolling stock stability. In fact, Nadal's equation allows to calculate the critical value of the derailment coefficient when the friction coefficient and the flange angle α vary.

By applying Nadal's equation two limit situations are identified:

- high safety level against derailment: $Q/P << 1$;
- low safety level against derailment: $Q/P > 1$.

Figure 4.6 shows the diagram of the derailment coefficient trend, which is obtained from Eq. (4.10) in function of the values of the flange angle α and coefficient f.

Studies of various train derailments have demonstrated that Eq. (4.10) can be simplified by the following expressions: $Q/P < 1.2$ for vehicle on axles and $Q/P < 1.3$ for vehicle on bogies [6].

Many railway companies carry out derailment risk analysis by considering a reference value V_R proportional to Nadal's derailment coefficient:

$$V_R = c \cdot \left(\frac{\tan \alpha - f}{1 + f \cdot \tan \alpha} \right) \tag{4.11}$$

where c is a safety coefficient (with $c < 1$, e.g. JR EAST "East Japan Railway Company" assumes $c = 0.15$).

Since the coefficient of friction and the ratios between the lateral and vertical components of the wheel-rail interaction force cannot be modified significantly, in order to improve safety levels, in the last few decades wheel profiles have been optimised with consequent increases in the flange angle α (up to 70° values). Clearly, the derailment risk increases when the ratio Q/P increases (see Eqs. (4.10) and (4.11)), as also confirmed by further investigations on train derailments, whose results are summed up in Table 4.4 [7].

Nadal's theoretical model underestimates the derailment risk. Indeed, in real cases the ratio between the static forces Q and P keeps constant in time intervals of the order of 0.2 –0.4 s [7], during which the train drives along track sections that can have irregularities with consequent vertical wheel accelerations (neglected in the model) and therefore with an increase in the derailment risk due to P value oscillations.

In any case, the safety assessment of the derailment risk is made in the most critical conditions, generally referred to the outer rail of the circular curves. The assessment of the maximum vertical load $Q_{e,max}$ and the maximum lateral force $Y_{e,max}$ acting on the outer rail can be made with the criteria illustrated in Chap. 5, by additionally taking into account the dynamic loads.

As a consequence of traffic loads accumulated over time, the angle α can decrease and lead—the other factors being equal—to the increase of derailment

Table 4.4 Q/P ratio and derailment risk class description ($\alpha = 60°$, $f = 0.2$)

Q/P	Risk class description
0.68	The line of action of the interaction force F is inclined around 34° and passes to the outer rail foot at a distance equal to $e = h_r \cdot 0.68 - \left(\frac{b_h}{2} + \frac{b_f}{2}\right)$. In case of UIC 60 rail (see Table 4.5) and e = 5 mm, if the rail weren't fixed to sleepers, it would overturn. This situation indicates instability
0.75	A wheel with a worn-out flange is likely to climb over a worn rail, while this is very unlikely in case of new or unworn wheel and rail
0.83	In a curve the wheel could climb over the outer rail
1.29	The derailment takes place independently of the geometry and conditions of the wheel and rail

Table 4.5 Characteristic values of the geometric parameters of the most common rails to use in the expression of Table 4.4 [2]

Rail section geometry	S41	S49	NP46	UIC54	UIC60	Ri60
Height h_r [mm]	138	149	142	159	172	180
Head width b_h [mm]	67	67	72	70	72	113
Foot width b_f [mm]	125	125	120	140	150	180
Area [cm^2]	52.7	53	59.3	69.3	76.9	77.1
Mass/meter [kg/m]	41.3	49.4	46.6	54.4	60.3	60.5
Inertia moment I_y [cm^4]	1368	1819	1605	2346	3055	3334
Inertia moment I_z [cm^4]	276	320	310	418	513	884
Section modulus W_{yh} [cm^3]	196	240	224	279	336	387
Section modulus W_{yf} [cm^3]	200.5	248	228	313	377	355
Section modulus W_z [cm^3]	44.2	51.2	52	60	68	135

risk (see Eq. (4.10)). Therefore, the rail profile is constantly monitored with precise measurements of the wear of the rail head (at 45°, vertical and horizontal, see Chap. 6).

References

1. D'Agostino V (2013) Fundamentals of mechanics applied to machines (in Italian, *Fondamenti di Meccanica applicata alle Macchine*), Maggioli Editore
2. Esveld C (2001) In: Modern railway track. 2nd edn, MRT-Productions
3. Pau M, Corona G (1998) Experiences on the contact area between wheel and rail based on the use of ultrasonic waves. Ingegneria Ferroviaria, 11
4. Egana JI, Vinolas J, Gil-Negrete N (2005) Effect of liquid high positive friction (HPF) modifier on wheel-rail contact and rail corrugation. Tribol Int 38:769–774
5. Eadie DT, Kalousek J, Chiddick KC (2002)The role of high positive friction (HPF) modifier in the control of short pitch corrugations and related phenomena. Wear 253185–192
6. Profillidis VA (2022) Railway planning, management, and engineering. (Fifth Edition), Routledge
7. Hay William (1982) Railroad Engineering. Wiley

Chapter 5
Introduction to Railway Track Design

Abstract In railway tracks the stresses and deformations induced by their own weight are negligible if compared to those deriving from thermal and vehicle loads. The latter can be distinguished into: quasi-static loads (weight force, centrifugal force and wind-induced force); dynamic loads, imputable to the effect of geometrical irregularities of the track and the rolling surface (wear, joints, points and crossings, etc.) or to defects on the surface of wheel rims. The ballasted tracks are designed by imposing that the deflection in the subgrade is always in an elastic or plastic field. The subsidences occurred during the track service life must remain below given threshold values in order not to jeopardise functionality and safety of the railway line. The criteria for calculating ballasted tracks are synthetically examined in this chapter and are based on the procedure described by Esveld [1]; for more detail, the interested reader may refer to [2, 3, 4].

5.1 Track Loads

5.1.1 Axle Loads

The values of static vertical loads of some rolling stock types are given in Table 5.1. With the same traffic flows, when the axle load value increases, track defects increase over time. This is another reason why the UIC (International Union of Railways) has classified the railway lines in function of the nominal axle loads applied to the track (see Table 5.2). The Italian railway infrastructure manager (RFI) plans and carries out monitoring the state of railway lines (see Chap. 6) at regular time intervals correlated to their importance and, thus, to their rail traffic intensity in terms of daily tonnage.

5.1.2 The Equivalent Load

The intensity of rail traffic expressed in daily equivalent tonnage is a specific traffic variable introduced by the UIC and used in railway engineering.

© The Author(s), under exclusive license to Springer Nature Switzerland AG 2023
M. Guerrieri, *Fundamentals of Railway Design*, Springer Tracts in Civil Engineering,
https://doi.org/10.1007/978-3-031-24030-0_5

Table 5.1 Number of axles and weight per axle of several rolling stock types [1]

Rolling stock type	Number of axles	Empty (kN)	Loaded (kN)
Trams	4	50	70
Light-rails	4	80	100
Passenger coaches	4	100	120
Locomotives	4–6	215	–
Freight wagon	2	120	225
Heavy freight trains (USA, Australia)	2	120	250 –350

Table 5.2 UIC load classification

Category	Axle load (kN)	Weight/m (kN/m)
A	160	48
B1	180	50
B2	180	64
C2	200	64
C3	200	72
C4	200	80
D4	225	80

For design purposes, railway lines can be classified into classes based on the equivalent load T_f as follows:

$$T_f = T_p \cdot \frac{V}{100} + T_g \cdot \frac{P_c}{18 \cdot D} \qquad (5.1)$$

where

- T_p: real load for daily passenger traffic [tonnes/day];
- T_g: rea l load for daily freight traffic [tonnes/day];
- V: maximum permissible speed [km/h];
- D: minimum wheel diameter of the most frequent vehicles in the line [m];
- P_c: maximum axle load with wheels of diameter D [tonnes].

Railway lines are also subdivided in function of the value of the equivalent load T_f observed (or scheduled) on at least 50 km long sections. Table 5.3 shows the Italian classification, also used in other European countries. In Italy the railway lines are usually dimensioned with equivalent loads ranging between $50,000 \leq T_f \leq 80,000$ tonnes/day [3].

Table 5.3 Italian railway lines classified based on equivalent loads

Group	Equivalent load T_f (tonnes/day)
1	$T_f > 102,000$
2	$70,000 < T_f \leq 102,000$
3	$40,000 < T_f \leq 70,000$
4	$25,000 < T_f \leq 40,000$
5	$12,500 < T_f \leq 25,000$
6	$6000 < T_f \leq 12,000$
7	$3000 < T_f \leq 6000$
8	$1000 < T_f \leq 3000$
9	$T_f \leq 1000$

5.1.3 Vertical Wheel Load

The total vertical wheel load Q_{Tot} transmitted from a generic wheel to the rail is the sum of two contributions: the quasi-static load Q_{qs} and the dynamic load Q_d [1]:

$$Q_{Tot} = Q_q + Q_d \tag{5.2}$$

$$Q_{qs} = Q_{st} + Q_c + Q_v \tag{5.3}$$

$$Q_{Tot} = (Q_{st} + Q_c + Q_v) + Q_d \tag{5.4}$$

where

- Q_{st} is the static wheel load, equal to the half of the axle static load measured in a straight horizontal railway line;
- Q_c is the increase in the wheel load on the outer rail in curves, in function of the non-compensated centrifugal force;
- Q_v is the increase in the wheel load on the outer rail in curves, in function of the transverse wind;
- Q_d is the dynamic wheel load component induced by sprung and unsprung masses, rail corrugations, welds and wheel flats [1].

Consider the forces acting on the vehicle of Fig. 5.1. By imposing the equilibrium to the rotation around the inner rail of the curve and by considering the meaning of Q_c and Q_v, it follows: $(Q_c + Q_v) \cdot s = (a_{nc} \cdot P/g) + F_v \cdot p_w$. Moreover, by employing Eq. (2.18) in Chap. 2, it results [1].

$$Q_c + Q_v = P \cdot \frac{p_c \cdot j}{s^2} + F_v \cdot \frac{p_w}{s} \tag{5.5}$$

Fig. 5.1 Quasi-static forces
acting on a vehicle in curve

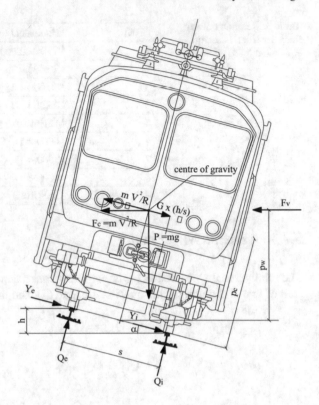

where

- $j = \frac{s \cdot v^2}{g \cdot R} - h$ is the superelevation deficiency;
- P is the vehicle weight per wheelset;
- F_v is the transversal wind force;
- s is the distance between the wheel-rail contact points of the two rails (1500 mm);
- v is the speed;
- g is the gravity acceleration;
- R is the curve radius;
- h is the superelevation;
- p_c is the distance of the vehicle centre of gravity from the plane passing through the rail head;
- p_w is the distance of the lateral wind force resultant from the plane passing through the rail head.

Q_c usually ranges between 10 and 25% of the static wheel load [1]. On the other hand, the maximum wheel load on the outer rail [1] (superelevation h > 0) is:

$$Q_{e,max} \approx \frac{1}{2}P + P \cdot \frac{p_c \cdot j}{s^2} + F_v \cdot \frac{p_w}{s} \tag{5.6}$$

For the sake of track dimensioning, the dynamic load can be assessed by multiplying $Q_{e,max}$ by the dynamic amplification factor (DAF); its analytical expressions are given in this chapter. As previously said, the dynamic forces are influenced by the geometric characteristics of the track: if the track is in good condition and rails do not show significant alterations, the values of dynamic impulsive-type actions are not a cause for serious concern. Vice versa, if the track geometry is poor and the rails are worn-out, the intensity of the dynamic forces are not negligible.

Also faults in railroad wheel shape (e.g. spalling defects) contribute to deteriorate the railway track; this is also the case for rail welds carried out imperfectly since, in this case, the dynamic load can overcome four times the corresponding static load Q, already at the speed of 120 km/h [1].

In short, irregularities on wheel and rail surfaces and poorly made welding cause a considerable increase in vertical actions as well as a rapid deterioration in the track regularity and, thus, poor comfort conditions for passengers.

5.1.4 Lateral Forces Acting on Rails

The total horizontal lateral force Y_{Tot} on the outer rail of the flat framework is given from the sum of two contributions: the quasi-static Y_{qs} and the dynamic forces Y_d [1]:

$$Y_{Tot} = Y_{qs} + Y_d \qquad (5.7)$$

$$Y_{qs} = Y_f + Y_c + Y_v \qquad (5.8)$$

$$Y_{Tot} = (Y_f + Y_c + Y_v) + Y_d \qquad (5.9)$$

where

- Y_f is the lateral force in curve caused by the impact of the wheel flange on the outer rail;
- Y_c is the lateral force due to the non-compensated component of the centrifugal force;
- Y_v is the component along the plane passing through the rail heads of the transversal force induced by the wind;
- Y_d is the lateral component of the dynamic force.

If we ideally assume that Y_c and Y_v entirely act on the outer rail, from the equilibrium of the forces along the direction of the plane connecting the rail heads (see Fig. 5.1 [1]), it results $Y_{e,max} \approx a_{nc} \cdot P/g + F_v$, and thus:

$$Y_{e,max} \approx P \cdot \frac{j}{s} + F_v \qquad (5.10)$$

Equation (5.10) is only a theoretical estimation of $Y_{e,max}$, since in real cases this force is also influenced by the wheelset type, transversal positions assumed by a vehicle while travelling along curves and adhesion forces.

5.2 Lateral Force Acting on the Flat Framework of the Track

The total lateral force acting on the flat framework (see Chap. 3) can be assessed by multiplying $Y_{e,max}$ by the dynamic amplification factor (DAF) [1]:

$$H = DAF \cdot \left(P \cdot \frac{j}{s} + F_v \right) \tag{5.11}$$

The lateral resistance of the flat framework of the track against the exerted lateral forces is provided as a result of the rail bending stiffness on their weak axes, torsional resistance of the fastenings and the sleeper lateral resistance in contact with the ballast layer.

Since the track has a limited resistance to lateral actions, it follows that elevated H values can cause transversal movements with consequent track deformations.

The minimum value of the lateral force H_{tr} (expressed in kN) against which the track must resist in order to guarantee stability, can be assessed with Prud'homme and Weber's formula [5, 6]:

$$H_{tr} < \alpha + \beta \cdot P \tag{5.12}$$

where P denotes the axle load (kN). The coefficients α and β of the expression (5.12) depend on the track conditions: $\alpha = 10$ and $\beta = 1/3$ for deteriorated or non-properly compacted ballast, $\alpha = 15$ and $\beta = 1/3$ for ballast in good condition and concrete sleepers.

5.2.1 Longitudinal Forces on the Track

Along the longitudinal direction the rails are subject to:

- compressive or tensile stresses due to change in temperature (thermal loads) and shrinkage stresses caused by rail welding
- forces from train wheels in acceleration and braking phases.

The temperature force can be estimated as follows. By denoting with L the unde-formed (original) rail length and with α the linear expansion coefficient of rail steel, the variation in rail length ΔL, due to the temperature variation ΔT (being

$\Delta T = T_{actual} - T_{initial}$), is given from:

$$\Delta L = \alpha \cdot L \cdot \Delta T \tag{5.13}$$

In a railway track without joints (i.e. in continuous welded rails) the longitudinal expansion of the rail is impeded, therefore by using Hooke's law (elastic region), it results:

$$\alpha \cdot L \cdot \Delta T - \frac{N \cdot L}{E \cdot A} = 0 \tag{5.14}$$

where N denotes the normal force, A the area of the rail cross section (see Table 4.5, Chap. 4) and E the Young modulus

Equation (5.14) clarifies that an increase in temperature results in a compressive normal force. From Eq. (5.14), the normal stress σ can be calculated as follows:

$$\sigma = \frac{N}{A} = \alpha \cdot E \cdot \Delta T \tag{5.15}$$

The longitudinal force due to the train acceleration or braking phases is not negligible with regard to the thermal loads, in that it assumes values up to 55 kN [7]. The braking force F_f can be especially assessed with the following relations in function of the axle load P [7]:

- $F_f = (0.12 - 15) \cdot P$ for electric locomotives;
- $F_f = 0.18 \cdot P$ for diesel locomotives;
- $F_f = 0.25 \cdot P$ for two-axle freight wagons.

5.3 Dimensioning Criteria with Static Analysis

The analysis is carried out by referring to quasi-static loads and hypothesising that rails behave like elastically supported beams.

5.3.1 Track Design in Case of Discrete Rail Support

The vertical force Q, transmitted from the wheel to the rail and acting in a certain time instant t, is distributed on n sleepers into n forces with value $F(x_i)$ each; therefore, it results:

$$Q = \sum_{i=1}^{n} F(x_i) \tag{5.16}$$

By hypothesising a Winkler behaviour of the support (sleepers) and by denoting with $w(x_i)$ the deflection of every spring (see Fig. 5.2), it follows [1]:

$$F(x_i) = C \cdot A_{rt} \cdot w(x_i) = k_d \cdot w(x_i) \tag{5.17}$$

Since A_{rt} is the contact area between rail and sleeper (rail support area) and $k_d = C \cdot A_{rt}$, the spring constant in a railway track with a homogeneous support can be indirectly obtained from the following equilibrium condition [1]:

$$k_d = \frac{\sum F}{\sum w} = \frac{Q}{\sum w} \tag{5.18}$$

in which $\sum w$ is the sum of all the significant deflections of sleepers near the point where the load Q is applied. Therefore, in case of more vertical forces Q_i applied at the same instant, k_d is obtained with the relation [1]:

$$k_d = \frac{\sum Q_i}{\sum w_j} \tag{5.19}$$

In fact, it results that [1]:

$$\Sigma Q_{ij} = \Sigma F_j = k_d \Sigma w_j \tag{5.20}$$

$$k_d = \frac{\sum Q_i}{\sum w_j}$$

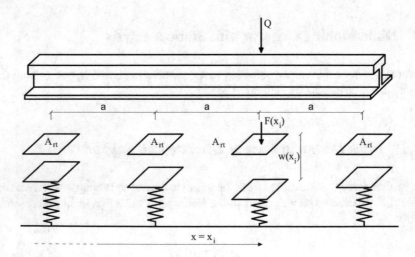

Fig. 5.2 Discrete elastic support model of a rail [1]

The spring constant k_d is exclusively a property of the support. Instead, the total spring constant of the track $k_{tot}(k_{tot} > k_d)$ depends on the track geometric configuration, and especially on the sleeper width and spacing and railway track support [1]. K_{tot} can be determined by the value of the maximum deflection w_{max} once the vertical load Q is exerted:

$$k_{tot} = \frac{Q}{w_{max}} \tag{5.21}$$

$$k_d = k_{tot} \frac{w_{max}}{\sum w}$$

The mean value of the contact pressure (cf. Figure 5.2) is [1]:

$$\sigma_{rt}(x_i) = \frac{F(xi)}{A_{rt}} \tag{5.22}$$

5.3.2 Track Design in Case of Continuous Rail Support

This is the case when distributed load p(x) acts with:

$$p(x) = k \cdot w(x) \tag{5.23}$$

where k is the foundation coefficient (i.e. the spring constant per unit length (N/m/m)).

By denoting the width of the supporting strip under the rail with b_t, the contact pressure on the continuous rail support is obtained as follows [1]: Fig. 5.3

$$\sigma_{rt}(x) = \frac{p(x)}{b_t} \tag{5.24}$$

In order to simplify the continuous rail support model with respect to the discrete model described in the previous section, the following equivalence condition must be imposed [1]:

$$k \approx \frac{k_d}{a} \tag{5.25}$$

where a is the spacing between the centres of discrete supports, corresponding to the sleeper spacing (see Fig. 5.2).

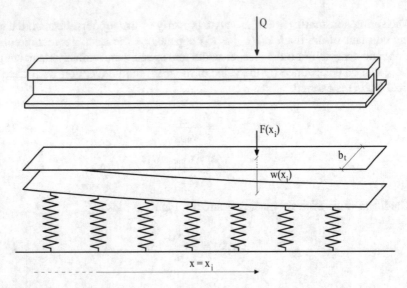

Fig. 5.3 Continuous elastic support model [1]

5.3.3   *Elastic Beam on an Elastic Foundation Model*

According to Zimmermann's model, the rail is considered to be infinitely long (e.g. continuous welded rail), with bending stiffness equal to EI. Moreover, the rail continuously rests on an elastic foundation modelled as a homogeneous elastic half-space with a foundation coefficient k. The wheel load Q is applied at the abscissa point x = 0 (see Fig. 5.4) and the rail deflection at the abscissa x_i is $w(x_i)$.

In order to determine the rail deflection $w(x)$, it is necessary to clarify the equilibrium conditions of an infinitesimal rail element (see Fig. 5.5) or, in other words, to satisfy the following conditions [1]:

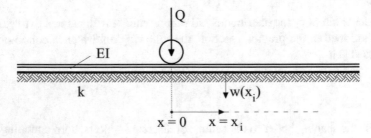

Fig. 5.4 Infinite elastic beam on an elastic foundation model [1]

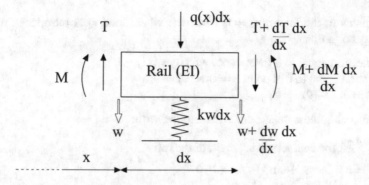

Fig. 5.5 Beam element model [1]

- *Equilibrium to the vertical translation*

$$q\,dx + \frac{dT}{dx}dx = k\,w\,dx \tag{5.26}$$

- *Rotation of the element $x + dx/2$ with regard to the vertical axis*

$$M + T\frac{dx}{2} + \left(T + \frac{dT}{dx}dx\right)\frac{dx}{2} - \left(M + \frac{dM}{dx}dx\right) = 0 \tag{5.27}$$

when neglecting the second-order differential term $\frac{dT}{dx}dx^2/2$, it follows:

$$T\,dx = \frac{dM}{dx}dx \tag{5.28}$$

The differential equation of the elastic line is:

$$\frac{d^2w}{dx^2} = -\frac{M}{EI} \tag{5.29}$$

where EI is the bending stiffness of the rail (expressed in (N/m^2)).

The previous expressions lead to the differential equation that determine deflections and stresses relative to given boundary conditions:

$$EI\frac{d^4w}{dx^4} + kw = q(x) \tag{5.30}$$

Or, by using a synthetic notation:

$$EI\,w^{IV} + kw = q(x) \tag{5.31}$$

In this case, the load $q(x) = 0$ while the wheel load Q is introduced into the boundary conditions:

- a deflection equal to 0 for $x = \infty$: $w(\infty) = 0$;
- a rotation equal to 0 at point $x = 0$: $w'(0) = 0$;
- for $x = 0$: $w'''(0) = \frac{Q}{2EI}$

By imposing these conditions and denoting with:

- $L = \sqrt[4]{\frac{4EI}{k}}$ the characteristic length of the rail
- $\eta(x) = e^{\frac{-x}{L}}\left(\cos\frac{x}{L} + \sin\frac{x}{L}\right)$ for $x \geq 0$
- $\mu(x) = e^{\frac{-x}{L}}\left(\cos\frac{x}{L} - \sin\frac{x}{L}\right)$ for $x \geq 0$

it follows [1]:

$$w(x) = \frac{QL^3}{8EI}\eta(x) = \frac{Q}{2kL}\eta(x) \tag{5.32}$$

The distributed reaction force of the foundation $p(x)$ is assessed by multiplying the corresponding deflection values $w(x)$ by k:

$$p(x) = k\,w(x) = \frac{Q}{2L}\eta(x) \tag{5.33}$$

The bending moment is given from [1]:

$$M(x) = \frac{QL}{4}\mu(x) \tag{5.34}$$

The expressions of $\eta(x)$ and $\mu(x)$, which determine the form of the elastic line, can also be employed to approximate the case of a finite rail length, on condition that the length is higher than the value $2\pi L$.

For an infinitely rigid beam ($EI = \infty$) of length 2L on an elastic foundation, at point $x = 0$, it results:

$$w_0 = \frac{Q}{2kL} \tag{5.35}$$

$$p_0 = k \cdot w_0 \tag{5.36}$$

$$M_0 = \frac{QL}{4} \tag{5.37}$$

5.3.4 Concomitant Action of Several Wheel Loads

Consider now the real case characterised by more wheels acting on the rail at the same time. Be Q_i the force which every wheel transmits to the rail. In order to solve the problem, it is possible to refer to the case previously examined (with only one wheel load) and apply the principle of the superposition of effects. Thus, at the abscissa point $x = 0$, it yields as follows:

$$w_0 = \frac{1}{2kL} \sum_i Q_i \cdot \eta(d_i) \qquad (5.38)$$

$$p_0 = k \cdot w_0 \qquad (5.39)$$

$$M_0 = \frac{L}{4} \sum_i Q_i \cdot \mu(d_i) \qquad (5.40)$$

where d_i is the distance of the load Q_i from the origin of the reference system ($x = 0$).

5.3.5 The Jointed Rails

A lot of railway lines have still the jointed rails with consequent breaks in the continuity of the rail (see Chap. 3). For this configuration it is possible to adopt the model of a hinged beam (at the hinge section the bending moment is zero) on an elastic foundation.

Should a reference system with abscissa $x = 0$ at the rail junction be chosen, the boundary conditions for the part of rail with $x > 0$ are the following:

- deflection equal to 0 for $x = \infty$: $w(\infty) = 0$
- bending moment equal to 0 at the point $x = 0$: $w''(0) = 0$
- and for $x = 0$: $w'''(0) = \frac{Q}{2EI}$

By introducing such conditions in the differential equation of the elastic line, it yields [1]:

$$w(x) = \frac{QL^3}{4EI} e^{\frac{-x}{L}} \cos\frac{x}{L} = \frac{Q}{kL} e^{\frac{-x}{L}} \cos\frac{x}{L} \qquad (5.41)$$

The absolute maximum bending moment occurs at the abscissa $x = \pi L/4$ [1]:

$$M_{max} = \frac{1}{4}\sqrt{2} \cdot e^{-\frac{\pi}{4}} \cdot QL \qquad (5.42)$$

5.4   The Dynamic Amplification Factor

Among the models mostly used for estimating the dynamic amplification factor (DAF), the following Eisenmann's [1] model considers the train speed V, track quality (by means of a coefficient φ) and the normalised variable z of Gaussian distribution:

$$DAF = 1 + z \cdot \phi \quad \text{for } V \le 60 \text{ Km/h} \tag{5.43}$$

$$DAF = 1 + z \cdot \phi \cdot \left(1 + \frac{V - 60}{140}\right) \quad \text{for } 60 < V \le 200 \text{ Km/h} \tag{5.44}$$

Statistically, assuming z = 1 means to accept the probability that the real stress values are higher than the calculation values in 31.7% cases. For z = 2 the probability is 4.6% while for z = 3 the probability is 0.3%.

At the design stage it is advisable to choose z = 3 in order to guarantee a higher safety standard and consequently a longer track lifecycle.

Table 5.4 shows the z values to choose for each track component.

For the flat framework of the track (rails, sleepers, fastenings) the value z = 3 is always advisable.

Table 5.5 shows the values to assign to the coefficient φ in case of tracks under good, average and bad conditions, respectively [1].

Other relations for calculating the DAF are [3]:

- Talbot's relation

$$DAF = 1 + 0.00062(V - 10) \tag{5.45}$$

- Relation used for the metro in Washington

Table 5.4   Values of the variable z

Probability (%)	z	Component of the railway track
68.3	1	Subgrade
95.4	2	Ballast
99.7	3	Rails, sleepers, fastenings

Table 5.5   Values of the coefficient φ

Track conditions	Coefficient φ
Good or very good	0.1
Average	0.2
Bad	0.3

$$DAF = \left(1 + 0.000259V^2\right)^{2/3} \tag{5.46}$$

- Empirical relation N. 1 [3]

$$DAF = 1 + V^2/45,000 \tag{5.47}$$

- Empirical relation N. 2 [3]

$$DAF = 1 + V^2/30,000 \quad \text{for } V \leq 100 \text{ km/h} \tag{5.48}$$

$$DAF = 1 + 4.5V^2/10^5 - 1.5V^2/10^7 \quad \text{for } V > 100 \text{ km/h} \tag{5.49}$$

5.5 Bending Stress in the Rail Foot Centre

In conformity with Eisenmann's model, as an effect of the dynamic actions on the rail, the expected maximum bending tensile stress in the rail foot centre due to the bending moment is [1]:

$$\sigma_{max} = DAF \cdot \sigma_{media} \tag{5.50}$$

with

$$\sigma_{media} = \frac{QL}{4W_{yf}} \tag{5.51}$$

where σ_{media} is the mean value of the rail bending stress at the centre of the rail foot, DAF is the dynamic amplification factor, Q is the load effectively transmitted from the wheel, $L = \sqrt[4]{\frac{4EI}{k}}$ is the characteristic length of the rail and W_{yf} is the section modulus of the rail foot (see Table 4.5 of Chap. 4).

For horizontal circular curves, Q is obtained by multiplying the nominal load (half of the wheelset weight measured in a straight section at stationary vehicle) by a multiplicative factor equal to 1.2 [1] in order to take into consideration the load increases due to the superelevation.

With the expressions reported above, it yields [1]:

$$\sigma_{media} = \frac{Q}{A} \frac{A\sqrt[4]{I}}{4W_{yf}} \sqrt[4]{\frac{4Ea}{k_d}}$$ (5.52)

where

- Q is the effective wheel load (kN);
- A is the surface of the rail cross section (m^2);
- I = I$_y$ is the moment of inertia of rail (m^4);
- W$_{yf}$ is the section modulus of rail relative to rail foot (m^3);
- E is the modulus of elasticity of rail steel (N/m^2);
- a is the sleeper spacing (m);
- k$_d$ is the spring constant of discrete support (N/m).

5.5.1 Stresses in the Rail Head

The vertical load Q generates in the rail head a semi-elliptical-shaped stress distribution as schematised in Fig. 5.6.

In agreement with Hertz's theory (see Chap. 4), the contact surface between the wheel and the rail is elliptical. Eisenmann devised a simplified calculation method of the shear stress distribution in the rail head based on the following assumptions:

- with wheels having a 60–20 cm diameter, two-dimensional calculation can be used;
- all the curvature radii of the bodies in contact are assumed to be infinitely large except that belonging to the wheel with a curve radius as r. This implies that the wheel/rail contact area is rectangular and the stress distribution has a semi-elliptical cylindrical form.

Fig. 5.6 Shear stress distribution in the rail head [1]

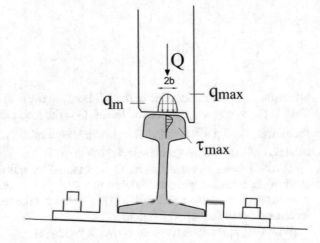

In this hypothesis the mean contact stress value q_m (expressed in kN) is [1]:

$$q_m = \sqrt{\frac{\pi EQ}{64(1 - v^2)rb}} \qquad (5.53)$$

In Eq. (5.53) the wheel radius is denoted with r, the width of the wheel/rail contact area with 2b (Fig. 5.6), the modulus of elasticity with E, Poisson's ratio with v. By assuming $E = 210,000$ N/mm², $v = 0.3$ and also b = 6 mm for steel (cf. values of the contact area in Table 4.1 of Chap. 4) and expressing Q in kN and r in mm, it yields [1]:

$$q_m = 1374\sqrt{\frac{Q}{r}} \qquad (5.54)$$

Once known the mean value of the contact stress q_m, it is possible to determine the maximum shear stress in the head of the rail [1]:

$$\tau_{max} \cong 0.3 \cdot q_m \qquad (5.55)$$

or,

$$\tau_{max} = 412\sqrt{\frac{Q}{r}} \qquad (5.56)$$

This simple model allows calculating the maximum shear stress at a depth of 4–6 mm below the rail head surface, which is a highly critical area because of the frequently observed defects like the so-called shelling [1].

The maximum shear stress must be compared with the permissible shear stress τ^* obtained with the Von Mises criterion [1]:

$$\tau* = \frac{\sigma^*}{\sqrt{3}} \qquad (5.57)$$

Since loads are repeated in time (fatigue), the permissible tangential stress can be estimated in function of the tensile breaking strength (σ_t):

$$\tau^* \cong 0.3\sigma^* \qquad (5.58)$$

From the previous relations (5.56) and (5.58) the expression for calculating the permissible effective wheel load Q* can be obtained. Therefore Q* is the maximum wheel load that does not produce rail breaks due to fatigue [1]:

$$Q^* = 49 \cdot 10^{-7} \cdot r \cdot \sigma^{*2} \qquad (5.59)$$

Table 5.6 Values of admissible shear stresses in the rail head

Steel type	Ultimate tensile strength σ^* (N/mm^2)	Admissible shear stress—incidental load (N/mm^2)	Admissible shear stress—repeated load (N/mm^2)
700	700	260	200
900	900	340	260

with r (mm), σ^*(N/mm^2) and Q* (kN).

Table 5.6 shows the values of ultimate tensile strength and the admissible shear stress which must not be overcome by incidental and repeated loads [1].

5.6 Sleeper Stresses

The estimation of the stress and deformation status of the sleepers can be carried out by using the free body diagram schematised in Fig. 5.7.

The maximum bearing force stressing every sleeper, due to the wheel load Q transmitted from the rail above, is:

$$F_{max} = DAF \cdot F_m \tag{5.60}$$

Considering the expressions:

$$\begin{cases} F(x_i) = CA_{rt}w(x_i) = k_d w(x_i) \\ k \approx \frac{k_d}{a} \\ w(x) = \frac{QL^3}{8EI}; \eta(x) = \frac{Q}{2KL}\eta(x) \\ L = \sqrt[4]{\frac{4EI}{K}} \end{cases} \tag{5.61}$$

It follows [1]:

$$F_m \approx \frac{Qa}{2L} = \frac{Q}{2}\sqrt[4]{\frac{k_d \cdot a^3}{4EI}} \tag{5.62}$$

Fig. 5.7 Contact stress distribution at the sleeper

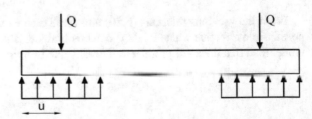

$$F_{max} = DAF \cdot \frac{Q}{2} \sqrt[4]{\frac{k_d \cdot a^3}{4EI}} \qquad (5.63)$$

where

- DAF is the dynamic amplification factor calculated for $z = 3$ (see Table 5.4);
- Q is the effective wheel load [kN];
- a is the sleeper spacing [m];
- k_d is the half support stiffness;
- EI is the single rail bending stiffness [Nm^2]

For the diagram in Fig. 5.7 the maximum moment is:

$$M_{max} = \frac{1}{4} \cdot F_{max} \cdot u \qquad (5.64)$$

The mean contact pressure between rail and sleeper on the most highly loaded sleepers can be determined with the expression [1]:

$$\sigma_{rt} = \frac{F_0 + F_{max}}{A_{rt}} \qquad (5.65)$$

where F_0 is the total prestressing force of fastening on rail support and A_{rt} denotes the effective rail support area of rail support. The following conditions must be met:

$$\sigma_{xt} \leq 1.0 \div 1.5 \ N/mm^2 \ for \ soft - wood \ sleepers \qquad (5.66)$$

$$\sigma_{rt} \leq 1.5 \div 2.5 \ N/mm^2 \ for \ hard - wood \ sleepers \qquad (5.67)$$

$$\sigma_{rt} \leq 4 \ N/mm^2 \ for \ concrete \ support \qquad (5.68)$$

5.7 Stresses on Ballast Bed and Formation

The ballast bed and the formation can be modelled as a two-layer system. The vertical stresses generated by vertical loads must be compared with the relative supporting capacities to avoid any uniform or differential deflections of the track [4].

The stresses on the two abovementioned layers can be calculated with Zimmermann's method, by estimating the dynamic amplification factor with Eisenmann's method. The maximum stress $\sigma_{sb,max}$ between sleeper and ballast bed exerted by the vertical load Q is [1]:

$$\sigma_{sb,max} = DAF \cdot \sigma_{sb,m} \qquad (5.69)$$

$$\sigma_{sb,m} = \frac{F_m}{A_{sb}} \approx \frac{Qa}{2L \cdot A_{sb}} = \frac{Q}{2A_{sb}} \sqrt[4]{\frac{k_d \cdot a^3}{4EI}} \tag{5.70}$$

The DAF related to the ballast can be calculated by assuming $z = 2$. In expressions (5.69) and (5.70) a denotes the sleeper spacing, A_{sb} the contact area between sleeper and ballast bed for the half sleeper, k_d the half support stiffness and EI the single rail bending stiffness [1].

The value of the permissible contact pressure on the ballast bed $\sigma_{sb,amm}$ is of the order of 0.50 N/mm^2 (it must be always $\sigma_{sb,max} \leq \sigma_{sb,amm}$).

The calculation of the maximum vertical stress on formation is made by super-imposing the load contributions due to single sleepers [1], the centres of which have coordinates x_i:

$$\sigma_i = \sigma_{max} \cdot \eta(x_i) \tag{5.71}$$

where

$$\sigma_{max} = DAF \cdot \frac{Qa}{2L \cdot A_{sb}} \tag{5.72}$$

$$\eta(x) = e^{\frac{-x_i}{L}} \left(\cos\frac{x_i}{L} + \sin\frac{x_i}{L} \right) \quad \text{for } x_i \geq 0 \tag{5.73}$$

In this case, the DAF must be calculated for $z = 1$.

5.7.1   Odemark's Method

Odemark's method allows the two layers under study (i.e. ballast bed and formation) to be analysed as a two-layer system with equivalent thickness H_e [1]:

$$H_e = 0.9 \cdot H \cdot \sqrt[3]{\frac{E_{ballast}}{E_{formation}}} \tag{5.74}$$

in which H is the real thickness of the ballast bed under the sleeper, $E_{ballast}$ is the modulus of elasticity of the ballast bed, $E_{formation}$ is the modulus of elasticity of the formation. If, for instance, the ratio between the two modules of elasticity is 3 and the ballast is 35 cm thick, the equivalent ballast thickness is $H_e = 45$ cm.

The vertical stress at a generic point at depth H_e (cf. Figure 5.8) is [1]:

$$\sigma_{z,i} = \sigma_i \cdot f(x_i) \tag{5.75}$$

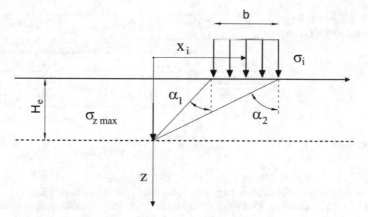

Fig. 5.8 Evaluation diagram of stress on ballast bed and formation

with [1]:

$$f(x_i) = \frac{1}{\pi}[\alpha_1 - \alpha_2 + 1/2(\sin 2\alpha_1 - \sin 2\alpha_2)] \tag{5.76}$$

$$\alpha_1 = \arctan \frac{x_i + b/2}{H_e} \tag{5.77}$$

$$\alpha_2 = \arctan \frac{x_i - b/2}{H_e} \tag{5.78}$$

It clearly must be [1]:

$$\sigma_{z,max} = \sum_i \sigma_{z,i} \tag{5.79}$$

The sleepers to consider in the calculation are the nearest to the point where the stress is to be estimated, in that the contribution of the other sleepers is less significant while their distance increases from that point. The determined value is to be compared with that one permissible σ_z^*, which can be assessed, for instance, with Heukelom and Klomp's empirical formula [8]:

$$\sigma*_z = \frac{0.006 \cdot E_{din}}{1 + 0.7 \cdot \log N} \tag{5.80}$$

where $E_{din} = (1, 2 \div 2, 5) \cdot E_{v2}$.

E_{v2} denotes the modulus of elasticity taken from the second load step in a plate loading test [8] and N is the number of load cycles. The characteristic values of E_{v2} and N are given in Table 5.7 [1].

Table 5.7 Characteristic values of E_{v2} and σ_z^* [1]

Classification	E_{v2} (N/mm^2)	σ_z^* (N/mm^2) $N = 2 \cdot 10^6$
Poor	10–20	0.011 –0.022
Moderate	50	0.055
Good	80–100	0.089 –0.111

References

1. Esveld C (2001) Modern railway track. 2 ed. MRT-Productions
2. Mundrey JS (2009) Railway track engineering. McGraw-Hill Education (India)
3. Bono G, Focacci C, Lanni S (1997) Railway track (in Italian, *La sovrastruttura ferroviaria*). CIFI
4. Li D, Hyslip J, Sussmann T, Chrismer S (2016) Railway geotechnics. Taylor & Francis
5. Prud'homme A, Weber O (1973) Technical aspects of high-speed trains. Track and its infrastructure. Electric installations1, 2 Travaux, pp 26–46
6. Esmaeili M, Hosseini SAS, Sharavi M (2016) Experimental assessment of dynamic lateral resistance of railway concrete sleeper. Soil Dyn Earthq Eng 82:40–54
7. Lichtberger B (2010) Track compendium. Eurail Press
8. Eisenmann J (1978) Railroad track structure for high-speed lines. In: Conference: "Railroad track mechanics & technology proceedings". Pergamon Press

Chapter 6
Railway Track Deterioration and Monitoring

Abstract The railway track is composed of several components as described in Chap. 3. Each component is subject to different types of horizontal and vertical loads (see Chap. 5) which affect degradation and failure processes. In order to guarantee reasonable safety and comfort levels and prevent catastrophic accidents, it is of fundamental interest to analyse the track efficiency and especially the conformity to normative threshold values of several geometric parameters of the track. For these reasons, nowadays the auscultation of the railway track is carried out by diagnostic trains capable of travelling on all railway lines and performing a series of inspections to evaluate the efficiency of the railway track with high precision.

6.1 Track Geometry

Static and dynamic forces (horizontal and vertical forces, respectively), thermal loads, material ageing can cause different types of deterioration of the railway track. Therefore, it is necessary to monitor the track efficiency and especially the conformity to normative limits of the values assumed by the following geometric parameters which will be detailed in the sections below [1, 2]):

- gauge;
- alignment;
- longitudinal level;
- cross level;
- cross level deviation;
- superelevation deficiency;
- twist.

With specific regard to rail, fastening and sleeper deterioration, it is crucial to measure the following defects and failures:

- vertical rail wear;
- 45-degree rail wear;
- lateral rail wear;
- presence of holes and cracks;

© The Author(s), under exclusive license to Springer Nature Switzerland AG 2023   111
M. Guerrieri, *Fundamentals of Railway Design*, Springer Tracts in Civil Engineering,
https://doi.org/10.1007/978-3-031-24030-0_6

- alumino-thermic and flash-butt welding wear in the long welded rails
- fastening condition;
- sleeper wear.

Track deterioration can increase the derailment risk (see Chap. 4), therefore the railway companies adhere to their own rules to set the acceptance limit of the geometric parameter values, beyond which operational restrictions are imposed (from speed limits to the temporary line closure).

6.1.1   Gauge

The gauge has been already defined in Chap. 2, Sect. 2.1. During the service life of the track, the distance between rails can increase with regard to the initial values as a consequence of the cyclic lateral forces transmitted from the train wheels.

If the gauge increases over a reasonable safety threshold, it can originate the slide-up derailment or the climbing derailment (see Chap. 4) due to the impulsive lateral forces which develop during the periodical movement of the wheelset with respect to the rails (the Klingel movement).

The gauge decrease below the lowest normative values is equally dangerous since, in some conditions, the wheel flange could jump over the rail.

6.1.2   Alignment

The alignment (expressed in mm) is measured on the plane passing through the top surface of the two rails and represents the distance between the inner side of the railhead and a straight segment of a given length (generally equal to 10 m [1]), which links two equidistant points from the section to be estimated. The alignment must be measured on both rails, by means of low-performance systems (e.g. manual systems) or high-performance systems which involve the use of diagnostic trains (in this case the obtained values are with loaded rail).

6.1.3   Longitudinal Level

The longitudinal level (expressed in mm) is measured on the vertical longitudinal plane and represents the distance between the plane passing through the top surface of the two rails and a straight segment of a given length (generally equal to 10 m [1]) connecting two equidistant points from the section to be estimated. The longitudinal level can be calculated with low-performance systems (e.g. manual systems) or with diagnostic trains.

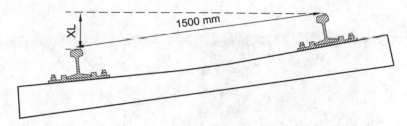

Fig. 6.1 Cross level

The localised rail deflection is frequently due to the ballast deterioration which, at the very worst, can be partly lacking below the sleepers.

6.1.4 Cross Level

The cross level XL, as previously written in Chap. 2, is the difference in height between the top surfaces of the rails; it coincides with the height of the minor cathetus of the rectangular triangle with a 1500 mm hypotenuse and a vertex angle equal to the angle between the running surface and the reference horizontal plane (see Fig. 6.1). It can be estimated with low-performance systems (calibres) or high-performance systems (diagnostic trains).

6.1.5 Cross Level Deviation

Cross level deviation XL_{dev} defines the difference (expressed in mm) between the cross level at a given point and the mean of the cross level values of two points set respectively 5 m before and after the point under consideration [1, 2]. By denoting the cross level at ith-point with XL_i, the cross level at $i + 1$-point set 5 m after the i-point with XL_{i+1} and the cross level at $i-1$-point set 5 m before the i-point with XL_{i-1}, it follows $XL_{dev} = XL_i - (XL_{i+1} + XL_{i-1})/2$.

6.1.6 Superelevation Deficiency

The superelevation deficiency (or cant deficiency) in a given track cross-section is the measure, expressed in millimetres, of the difference between the cross level XL and the design superelevation h (see Chap. 2).

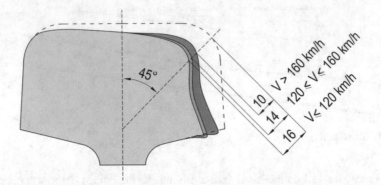

Fig. 6.2 Maximum permissible 45-degree wear according to Italian regulations [1]

6.1.7 Twist

The definition of twist γ can be found in Sect. 2.4 of Chap. 2. Also this geometric parameter can be measured with low- or high-performance systems.

The theoretical twist value is zero in straight line sections and equal to the gradient of the longitudinal superelevation of rails in transition curves. In this case the aim is to reduce as much difference as possible between the theoretical (design value) and actual twist values. This is so because, as observed in Chap. 2, in the sections with γ ≠ 0 the two rails are not coplanar; thus, one in the four wheels of vehicle bogies can turn out to be unloaded with consequent Q/P ratio increases (see Chap. 4) and consequent risk of wheel-climbing derailment.

6.1.8 Vertical Rail Wear

Rail is mainly affected by surface defects due to the interaction between wheel and rail [3]. The vertical wear is the loss of metal from the head of the rail profile (expressed in mm), compared to its theoretical profile, measured along the vertical axis of the cross section (Fig. 6.2).

6.1.9 45-Degree Rail Wear

The 45-degree wear is the loss of metal from the head of the rail compared to its theoretical profile, measured on a 45-degree-inclined line segment [2] (Fig. 6.2). The wear is detected with manual devices or with diagnostic trains.

6.1.10 Lateral Rail Wear

The lateral wear is the loss of metal from the head of the rail (expressed in mm) compared to the theoretical profile, measured along the straight line passing through the vertices of the fishing surfaces of the head of the rail [2].

6.1.11 Other Geometric Parameters

According to Italian regulations the following additional geometric parameters are measured for railway lines with speed $V > 250$ km/h:

- alignment on 20 m base "A_{20}";
- longitudinal level on 20 m base "L_{20}";
- cross level deviation on 20 m base XL_{dev-20};

which represent, respectively, the alignment, longitudinal and cross level deviation values, all obtained on a base of measurement of 20 m rather than 10 m.

6.1.12 Rail Defect Coding System

Rails can show more or less crucial surface defects due to construction errors or to fatigue or wear processes. In particular, wear is influenced by such factors as axle load, traffic density and speed, wheel and rail profile, material properties and hardness, curvature and superelevation and lubrication [3].

In Italy, rail defects were first codified in the *"Catalogue of the rail defects"* published in 1966 [4].

In UIC Code 712 R "Rail defects" [5] every defect is identified by a code composed of four digits:

- The first digit indicates:

  1. defects in rail ends;
  2. defects away from rail ends;
  3. defects resulting from damage to the rail;
  4. weld and resurfacing defects.

- The second digit indicates:

  - the place in the rail section where the defect originated;
  - the welding method, in the case of weld and resurfacing defects.

- The third digit indicates:

– the pattern of the defect in a broken or cracked rail;
– the nature of the defect in a damaged rail;
– the cause of the defect in a damaged rail.

- The fourth digit allows for a further classification based on the type of defect as and when required.

The general UIC coding system is given in Fig. 6.3 [5].

Figures 6.4, 6.5, 6.6, 6.7, 6.8, 6.9, 6.10 and 6.11 give a few coding examples, according to the UIC coding system [5].

The pitch corrugation—also named undulation—(Fig. 6.5) does not affect safety but it implies an increase in rolling noise and greater dynamic forces on vehicles and track. It can be classified [6] as follows:

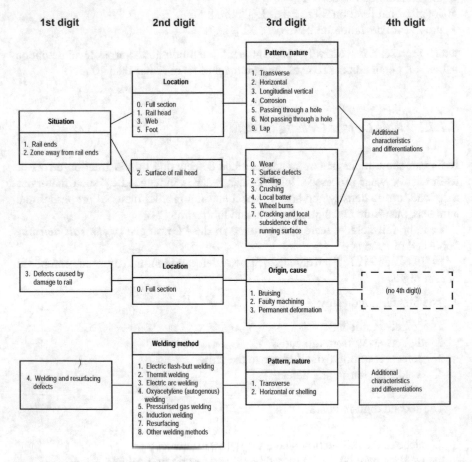

Fig. 6.3 UIC Rail defects coding system (*Source* [4])

Fig. 6.4 Longitudinal
vertical cracking (113)

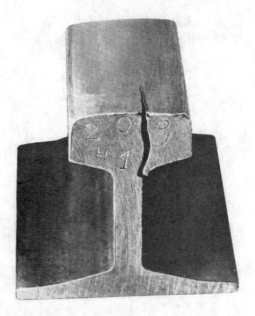

113

Fig. 6.5 Long pitch
corrugation (2202)

2202

- undulatory wear due to heavy traffic: observed in railway lines with heavy loads
  (up to 40 t for axle) and slow traffic. The wavelength is $\lambda = 200$ –300 mm.
- undulatory wear due to promiscuous-traffic on light rails: especially observed on
  rails with weight per unit length of 47–54 kg/m. The wavelengths range between
  500 –1500 mm.
- undulatory wear on the slab track and small radius curves: mostly observed in
  curves with a radius below 400 m. The wavelength is $\lambda = 45$ –60 mm.

Fig. 6.6 Abnormal vertical wear (2204)

2204

Fig. 6.7 Horizontal cracking at the web-head fillet radius (1321)

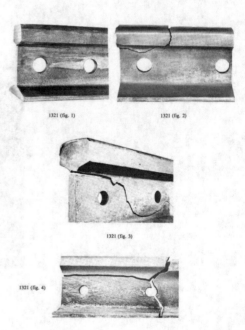

1321 (fig. 1)   1321 (fig. 2)

1321 (fig. 3)

1321 (fig. 4)

- wear on high-speed railway lines in straight sections: frequent on straight sections or curves with a very high radius (modest interactions between wheel flange and rail). The wavelength is around 25 –80 mm (Figs. 6.6–6.11)

Fig. 6.8 Longitudinal
vertical crack (piping) (233)

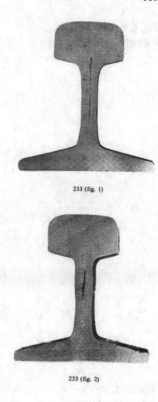

233 (fig. 1)

233 (fig. 2)

6.1.13 Fastening and Sleeper Examination

An indirect qualitative evaluation of the fastening resistance can be carried out by
analysing the values of the gauge depurated from the horizontal rail wear values.
By using a sight inspection or an automated video inspection [1] it is possible, on the
other hand, to control any coming out of the fastener bolts. Also the analysis of the
cracking status of reinforced-concrete sleepers is carried out with a sight inspection
or an automated video inspection.

6.2 Rail Corrugations

The formation of corrugations (undulatory wear) is a phenomenon extremely felt on
circular curves, especially those with a small radius. Numerous studies have shown
that corrugations are generated by the following main actions:

---

[1] The automated video inspection is carried out for example with the following diagnostic trains:
Euclide, *Galileo* and *Archimede*.

Fig. 6.9 Kidney-shape
fatigue crack (211)

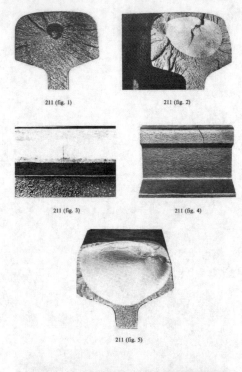

211 (fig. 1)   211 (fig. 2)

211 (fig. 3)   211 (fig. 4)

211 (fig. 5)

Fig. 6.10 Horizontal
cracking of the web

422.1

422.2

Fig. 6.11 Star-cracking of
fishbolt holes (135)

135 (fig. 1)

135 (fig. 2)

- excessive wheel/rail sliding caused by small-sized radii of horizontal curves;
- vertical load oscillations caused by rail and wheel irregularities;
- stick–slip between wheel and rail accompanied by vertical load oscillations.

The stick–slip actions especially cause the accelerated rail wear. Intensity and frequency of the sound waves emitted during the wheel motion are strictly linked to the characteristics of the micro- and macro-undulation of the rail surface (Fig. 6.12) and wheel irregularities.

A study carried out in Japan [7] showed, by means of a corrugation simulator, how rail corrugations originate and develop. Experimentation was carried out by analysing more than one condition:

- longitudinal creepage equal to 1, 2, 3, 4, 5%;
- lateral creepage: 0%;
- static load: 253 kN;
- wheel material: SMC 435 (harden);
- wheel tangential speed: 1570 mm/s (300 rpm).

The study showed the formation of two different types of irregularity:

- the former, characterised by a short pitch corrugation (1 kHz frequency) and wavelength of 1.6 mm;
- the latter, composed of long wave corrugation of 30 mm wavelength and about 50 Hz frequency.

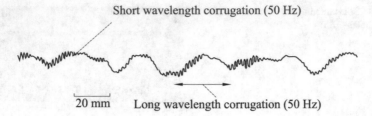

Fig. 6.12   Waveform of final corrugation profile under 1% creepage [7]

The conclusions of the study can be synthesised as follows [7]:

- the steady state longitudinal slip stimulates the development of corrugation type and wavelength;
- the short pitch corrugation is developed under low creepage conditions;
- the long wavelength corrugation is mainly developed under the relatively large slip condition.

6.3   The Rail Service Life

The rail service life is normally estimated in terms of cumulated traffic values (expressed in millions of tons) which a rail is expected to reach before its wear exceeds the threshold levels, thus requiring its replacement.

Experiments have shown [8] that the wear of the top surface of a rail, traffic conditions being equal, depends on the curvature of the railway line section.

Table 6.1 shows the values of the ratio "c" between the rail service life in curve sections and in straight sections, when the degree of curvature DC varies (with DC = 5730/R, where R denotes the curve radius). The effects of the curvature are negligible only for radius values over 11,000 m [7]. On curves with 1000 m radius, rail service life is equal to 30% of that typical of straight sections. The use of LCF lubricants (see Chap. 4) in a curve limits the service life deterioration. A study carried out by the American Railway Engineering Association in twenty railway lines [8] has shown that the vertical wear increases 0.76 every MGT (million gross tonnes) on average. More recently, in rails with a structural steel grade 900 A (see Chap. 3) such an increase has been observed to be between 1 and 2 mm every 100 MGT in the absence of LCF lubricants and 1 mm every 100 MGT in case of lubricated wheel flanges. [3, 10].

The lateral rail wear is mainly localised near the short radius curves and is nearly always greater than the vertical rail wear, but there is no direct proportionality between wear and cumulated traffic values. All the experiences made so far point out a strong correlation between the service life and mechanical resistance parameters for rails such as the weight per unit length, hardness (Brinell "BHN" or Vickers "HV") and traffic. Among the most well-known relations, the following is worth to be mentioned:

Table. 6.1 Reduction of the rail service life in curve sections

Degree of curvature DC	c = rail service life on curve sections/rail service life on straight sections	
	Lubricated rails	Non-lubricated rails
0.0 –0.5	1	1
0.5 –1.5	0.87	1
1.5–2.5	0.74	0.88
2.5 –3.5	0.61	0.79
3.5–4.5	0.49	0.7
4.5–5.5	0.38	0.62
5.5–6.5	0.3	0.55
6.5–7.5	0.22	0.48
7.5–8.5	0.16	0.44
8.5–9.5	0.12	0.4
> 9.5	0.1	0.37

$$L = K \cdot W \cdot D^{0.565} \tag{6.1}$$

where L denotes the life service of the rail (expressed in the tonnes cumulated during the rail service life), W is the rail weight per unit length expressed in pounds (1 pound = 0.454 kg) per yard (1 yard = 0.914 m), K denotes the coefficient based on the rail type (see Table 6.2), D is the yearly traffic load, expressed in million tonnes/year.

In order to determine the rail service life L, the first step is to set the replacement threshold, that is the value of the permissible vertical wear (UV_{amm}) beyond which the rail must be inevitably replaced. As previously pointed out, a rail made of 900 A steel causes a vertical wear always below 2 mm every 100 MGT ($\delta UV_{standard}$). It follows that the total number of tonnes cumulated during the service life results from the ratio between the permissible vertical wear UV_{amm} (expressed in mm) and the standard increase factor of the vertical wear $\delta UV_{standard}$ (mm/100 MGT).

The parameter K value depends on the rail type and on the rail connection technique. The long welded rails (LRS or CWR) have, on average, a longer service life than that characterising jointed rails, therefore K values are more elevated for the long welded rails (see Table 6.2).

Table. 6.2 K values

Weight per unit length (kg/m)	K	
	Jointed rail	Welded rail
55–60	0.9538	1.3544
> 60	0.9810	1.3930

In order to estimate K on the line sections at non-zero curvature it is necessary to proceed as follows:

- to determine the total length of straight sections S_T;
- to measure the length $S_{c,i}$ of each curve in the railway line under consideration and calculate the degree of curvature DC;
- to set the value of permissible vertical wear UV_{amm} (in mm), beyond which the rail must be replaced;
- to set the standard increase factor of the vertical wear $\delta UV_{standard}$ (mm/100 MGT) based on the rail type;
- to estimate the rail service life L (expressed in MGT) as if the railway line were entirely straight:

$$L = \frac{UV_{amm}}{\delta UV_{stantard}} \qquad (6.2)$$

- To calculate the equivalent K value:

$$K = \frac{L}{W \cdot D^{0.565}} \qquad (6.3)$$

- To calculate the virtual length of the line S_v with the following relation:

$$S_v = S_T \cdot \frac{L}{W \cdot D^{0.565}} \cdot c_{str} + \sum_i S_{c,i} \cdot \frac{L}{W \cdot D^{0.565}} \cdot c_i \qquad (6.4)$$

in which the coefficients "c" for straight lines (c_{str}) and for curves (c_i) are determined by using the values in Table 6.1.

- To calculate the ratio between the virtual length of the line S_v and the real length S_r, corresponding to the actual K value ($K = K_{act}$):

$$K_{act} = \frac{S_v}{S_r} = \frac{1}{S_r} \cdot \left[S_T \cdot \frac{L}{W \cdot D^{0.565}} \cdot c_{str} + \sum_i S_{c,i} \cdot \frac{L}{W \cdot D^{0.565}} \cdot c_i \right] \qquad (6.5)$$

- To assess the actual service life with the expression:

$$L_{act} = K_{eff} \cdot W \cdot D^{0.565} \qquad (6.6)$$

Finally, the rail service life, expressed in years, can be calculated by the ratio between the actual service life L^*_{act} (expressed in MGT) and the yearly traffic load D (expressed in MGT/year):

$$L^*_{act} = \frac{L_{eff}}{D} \tag{6.7}$$

6.4 The Track Geometric Quality Index

Isolated defects of a railway track can affect comfort and safety. By properly processing the different geometric parameter values, the track quality can be globally assessed with reference to the time instant when it was monitored. For each geometric parameter (gauge, alignment, twist, etc.) the standard deviation σ is calculated from the sample of values measured in a track section of given length, generally 100, 200 or 1000 m [1, 6, 11]:

$$\sigma = \sqrt{\frac{\sum_{i=1}^{N} (x_i - x_m)^2}{N}} \tag{6.8}$$

In which N stands for the size of the sample, x_i is the observed value of the geometric parameter under examination (e.g. one value per track metre) and x_m is the mean value of these observations.

By particularising Eq. (6.8) the standard deviation of the individual geometric parameters can be evaluated: for instance, alignment σ_a, longitudinal level σ_L, transversal level σ_{XL}, twist σ_γ gauge σ_S, etc.

The track geometric quality index TQI is defined from [9]:

$$TQI = \sqrt{w_a \cdot \sigma_a^2 + w_L \cdot \sigma_L^2 + w_{xL} \cdot \sigma_x^2 + w_\gamma \cdot \sigma_\gamma^2 + w_j \cdot \sigma_j^2 \ldots} \tag{6.9}$$

In Eq. (6.9) w_j denotes the weighting factor assigned to the geometric parameter j-th. The number of geometric parameters to introduce into Eq. (6.9) is variable and is fixed by the infrastructure manager of the railway network. The abovementioned standard deviations and the track geometric quality index TQI are compared with the corresponding threshold values established by the rules in each country. This comparison helps the infrastructure manager to plan maintenance tasks.

The Italian geometric standards for railway tracks [2] currently provide for three quality levels, which imply full functioning of the railway lines, and one level asking for traffic restrictions in terms of train speed reduction and traffic interruption:

- 1st quality level: the track geometry is very good and does not require any maintenance or renewal action;
- 2nd quality level: the geometry must be kept under control over time; it is necessary to evaluate the causes which have led to degradation and to estimate how quickly the defect will worsen in the time so as to plan the necessary maintenance interventions;
- 3rd quality level (or intervention level): maintenance works are required within established times;
- Level requiring traffic restrictions: train speed reductions or traffic interruptions are to be made.

The defects in the latter case, together with those of the 3rd level, are "significant defects" of the track geometry [1].

In 2001 [1] the Italian infrastructure manager RFI established two geometric quality indices: (1) track geometry quality index in a railway line (IQBT) and (2) track geometry quality index in a station (IQBS), by using the values of following defect indices which were calculated for subsequent sections, each 200 m long:

- defect index of the longitudinal level = standard deviation of the longitudinal level;
- defect index of the alignment = standard deviation of the alignment;
- defect index of the cross level = standard deviation of the cross level;
- overall defect index equal to, for every 200 m length, the highest value of the defect indices described above.

The track quality indices were defined by the Italian infrastructure manager RFI as follows [1]:

- *track geometry quality index in a railway line (IQBT)*: mean value of the overall defect index (one every 200 m) concerning the examined line;
- *track geometry quality index in a station (IQBS)*: mean value of the defect indices of the longitudinal level (one every 200 m) in the considered station.

Such indices had to be always inferior to given threshold limits based on the maximum rank speed (see Chap. 2) of the railway line under examination (Table 6.3).

6.4.1   Deterioration Curves of the Quality Indices

Numerous studies show that the deterioration curves of track geometric parameters or track geometric quality index TQI decrease over time as shown in Fig. 6.13 [3, 6,

Table. 6.3 Threshold limits of the track quality indices for railway lines (IQBT) and stations (IQBS) [1]

Railway track geometry quality index		
Maximum rank speed (Km/h)	IQBT (mm)	IQBS (mm)
V < 80	3.1	3.7
100 ≤ V < 100	2.8	3.4
100 ≤ V < 120	2.6	3.2
120 ≤ V < 140	2.4	3.0
140 ≤ V < 160	2.0	2.6
160 ≤ V < 180	1.9	2.5
180 ≤ V < 200	1.8	2.4
200 ≤ V < 250	1.4	2.0

11]. For merely descriptive purposes, it may be assumed that the deterioration model of the generic quality index Q(t) be of the type:

$$Q(t) = Q_i - e^{c_i \cdot t} \tag{6.10}$$

Since Q(t) is the value of the quality index at time t, Q_i is the initial quality of the new railway track, or consequent to the i-th maintenance action, and c_i denotes the deterioration rate coefficient.

A fairly reliable deterioration model permits to plan successful maintenance interventions (Fig. 6.13), optimise costs and reduce discomfort for travellers.

Table 6.4 summarised the maintenance actions required on the railway track elements during their life cycle.

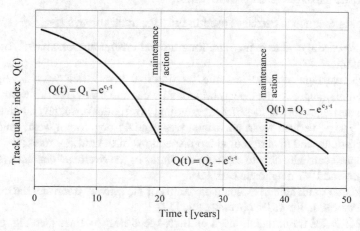

Fig. 6.13 Deterioration curves of the generic track quality index Q(t)

Table. 6.4 Typical
maintenance actions during
the railway track life cycle
(*Source* [3])

Actions on the track	Traffic load(MGT)	Frequency (years)
Tamping	40–70	4–5
Grinding	20–30	1–3
Ballast cleaning	150–300	12–15
Rail renewal	300–1000	10–15
Timber sleeper renewal	250–600	20–30
Concrete sleeper renewal	350–700	30–40
Fastening renewal	100–500	10–30
Ballast renewal	200–500	20–30
Formation renewal	>500	40

6.5   Diagnostic Devices

Railway track auscultation with diagnostic trains is carried out during the so-called test runs which can be:

- ordinary: planned and performed regularly and based on the railway line importance;
- extraordinary: performed in special circumstances, e.g. in the presence of track anomalies;
- operational: to verify and validate the precision of diagnostic devices installed in diagnostic carriages or diagnostic trains.

The main diagnostic carriages used in Italy for track surveys are:

- Cartesio carriage: able to measure longitudinal level, transversal level, twist and alignment;
- Euclide carriage: able to measure longitudinal level, transversal level, twist, alignment, gauge, vertical wear, horizontal wear and 45-degree rail wear;
- Talete carriage: able to measure longitudinal level, transversal level, twist, alignment, gauge, vertical wear, horizontal wear and 45-degree rail wear, train bogie accelerations, and to carry out video inspections of the railway track;
- Aldebaran carriage: able to evaluate the status of the overhead contact system;
- Caronte and PV7 diagnostic carriages;
- Archimede train: able to evaluate the status of the railway track and the overhead contact system, the GSM coverage etc. [12];
- DIA.MAN.TE train: mainly used on high-speed railway lines (see Fig. 6.14). It can survey over 200 parameters concerning the railway track, the overhead contact

Fig. 6.14 Diagnostic train DIA.MAN.TE

system, the signalling systems (e.g. ERTMS system) and telecommunication systems (e.g. GSM-R radio system).

References

1. Technical rules on railway track for railway line with speed ≤ 250 km/h. (code: RFI TCAR ST AR 01 001 A, RFI, 30/11/01). Italian RFI (in Italian)
2. Technical rules on railway track for railway line with speed < 300 km/h (code: RFI TCAR ST AR 01 001, D RFI, 31/21/2013). Italian RFI (in Italian)
3. Tzanakakis K (2013) The railway track and its long term behaviour. Springer
4. Catalogue of the rail defects (1996) Italian RFI (in Italian)
5. UIC (2002) Code 712 R "Rail defects", 4th ed
6. Lichtberger B (2010) Track compendium. Eurail Press
7. Suda Y et al (2002) Experimental study on mechanism of rail corrugation using corrugation simulator. Wear 253:162–171
8. Magee GM (1969) Proceedings of the AREA, vol 70 (Bulletin 615, September–October 1968), p 81
9. Bunjex JA (1959) Recente developments affecting rail section, "Report of rail Committee Assignment 9". In: Proceedings of the AREA, vol 60, p 971
10. Esveld C (2001) Modern railway track. 2nd ed. MRT-Productions
11. Veit P (2006) Qualität im Gleis–Luxus oder Notwendigkeit? EI—Eisenbahningenieur 57(12):32–37
12. Guerrieri M, Parla G, Ticali D (2012) A theoretical and experimental approach to reconstructing the transverse profile of worn-out rails. Ingegneria Ferroviaria 67(1):23–37

Chapter 7
Basics of Switches and Crossings

Abstract Switches and crossings are the essential elements of railway lines, stations and yards, in that they allow a train to change directions or a vehicle to be transferred from one track to another. Their main characteristics and classifications are concisely described in this chapter.

Switches and crossings are railway devices which connect two or more tracks, or allow tracks to cross one another [1, 2]. These devices allow trains to change direction from one track to another without stopping.

They are used at stations, freight terminals, marshalling yards, etc.

Switches take a large variety of configurations, but they can be subdivided into two principal forms and a third, combining the two [2]:

- simple or multiple turnouts, by which a track can be divided into two or three tracks;
- crossings, in order to simply cross two tracks intersecting one another at the same level;
- turnout crossings, which combine the characteristics of turnouts and crossings.

7.1 Turnouts

A simple turnout (Fig. 7.1) can be of two types: right-hand or left-hand depending on whether it diverts the trains to the right or to the left.

A simple turnout has a mobile part, i.e. the tongue frame (tapered movable rails), and a fixed part, called frog. The two parts are connected by means of four sections of intermediate rails. The most extreme ends of the mobile track (tongues), around 2 m long, show a tampered profile to make them fit with the stock rails (i.e. motionless rails). The tongues are linked to one another and are manoeuvred with an articulated quadrilateral system. The movement occurs with levers (manual operation) or with electric and hydro-pneumatic devices (automatic operation).

The tongue frame can have two distinct positions:

© The Author(s), under exclusive license to Springer Nature Switzerland AG 2023   131
M. Guerrieri, *Fundamentals of Railway Design*, Springer Tracts in Civil Engineering,
https://doi.org/10.1007/978-3-031-24030-0_7

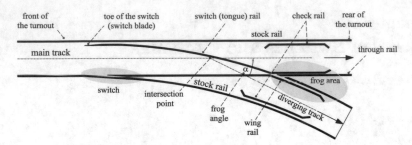

Fig. 7.1 Main constituents of a right-hand turnout

- continuity with the main track;
- continuity with the diverging track.

The rotations of the tongue frame occur on two hinges called "heels".

The direction of a turnout is known as the "facing direction" if a train approaching the turnout needs to face the thin end of the switch. The direction is known as "trailing direction" if the train crosses the switch in the trailing direction [1].

Turnouts have some discontinuity in rail sections: thus, the presence of some gaps, the so-called "dangerous spaces", prevent wheels from having any rail contact and require two checkrails to be inserted to avoid derailments.

The diverging track has a horizontal curve made of an arch of a circumference or a transition curve (cubic parabola or clothoid). In order to take the deviation angle into account in the turnout classification, the trigonometric tangent of the angle formed by the diverging track with regard to the main track is measured near the frog (the so-called frog angle).

The turnouts are identified with an alphanumeric code with five elements, for instance the Italian code S60/170/0.12d, with the following meaning [3]:

- 1st character (S) indicates the turnout type (S for simple);
- 2nd character (60) indicates the mass per unit length of rails forming the turnout;
- 3rd character (170) indicates the radius of the diverging track, expressed in metres;
- 4th character (0.12) indicates the tangent of the frog angle;
- 5th character (d) defines the direction of the diverging track: d for right, s for left.

For safety reasons, in turnouts the value of the track gauge is correlated to the radius R of the diverging track (Table 7.1).

Table. 7.1 Gauge values in turnouts

Radius R of the diverging track	Gauge
R ≥ 245 m	1435 mm
150 m ≤ R ≤ 244 m	1445 mm
110 m ≤ R ≤ 149 m	1460 mm
70 m ≤ R ≤ 109 m	1465 mm

Table. 7.2 Turnout types and permissible speeds

Turnout code	Radius (m)	V (km/h)	Frog	Exit angle
S60 UNI/170/0.12	170	30	Straight	6° 50′ 34″
S60 UNI/250/0.092	250	30	Straight	5° 15′ 30″
S60 UNI/250/0.12	250	30	Curved	6° 50′ 34″
S60 UNI/400/0.12	400	60	Straight	4° 13′ 46″
S60 UNI/400/0.074	400	60	Curved	5° 21′ 55″
S60 UNI/1200/0.040	1200	100	Straight	2° 17′ 26″
S60 UNI/3000-∞/0.022	3000	160	Straight	1° 15′ 23″
S60 UNI/6000-∞/0.015	6000	220	Straight	0° 51′ 26″

7.2 Speeds on Turnouts

All the turnout components (rails, tongues, frog, checkrails etc.) are placed with their axis in vertical (i.e. lack of 1/20 inclination generally provided in the other sections of the railway line) and on the same plane; therefore, while the train travels, there is no compensation of the centrifugal acceleration due to the superelevation. By imposing on the diverging track a non-compensated acceleration value $a_{nc} = 0.65$ m/s^2, it follows (see Chap. 2):

$$a_{nc} = V^2/(3.6^2 \cdot R) \tag{7.1}$$

hence

$$V = 3.6(0.65 \cdot R)^{1/2} = 2.91(R)^{1/2} \tag{7.2}$$

The maximum speed of the diverging track, moreover, depends on the speed standard permitted by the railway signalling rules. The turnout classification and the relevant allowed speeds are given in Table 7.2. If the diverging track is built with a transition curve, the centrifugal force ideally varies with continuity from the maximum entry value 0.65 m/s^2 up to the value 0 m/s^2 at the turnout exit.

7.3 Curve Turnouts, Diamond Crossings and Crossovers

The switches can be also placed in curve (Fig. 7.2). In this specific configuration, the diverging track (with a variable radius, measured with regard to the main track) can be internal or external to the correct track in curve. In any case, the radius of the diverging track must not be lower than 150. Generally, the switch radius ranges between 150–500 m with permissible speeds at the diverging track of 35–65 km/h [2].

Fig. 7.2 Examples of curve turnouts

Three fundamental configurations can be identified [4]:

1. diverging track turned towards the inner curve;
2. diverging track turned towards the outer curve;
3. diverging track turned towards the outer curve with a curvature of opposite sign compared to that of the direct branch.

The diamond crossing is used if two tracks of the same or different gauges cross each other. The configuration can be symmetrical or asymmetrical. The diamond crossing does not have a tongue frame and does not allow any deviations from the vehicle travelling direction. The classification is made by means of an alphanumeric code with three characters; for instance, the Italian code is as follows:

- 1st character, "I" stands for intersection;
- 2nd character: indicates the mass per unit length of the rails;
- 3rd character: indicates the value of the tangent of the deviation angle (exit angle).

By way of an example, a diamond crossing with UIC60 rails between two intersecting tracks with an angle of $6° 50' 34''$ is identified with the code I60/0.12.

In some cases, slip arrangements can be provided to allow vehicles to change tracks in one direction (single slip) or in both directions (double slip). A double slip (Fig. 7.3) has two simple switches, one right and one left, which follow one another at a short distance, partially embedded to save space: the frog of the former switch is near the head of the latter switch.

The need to reduce the dangerous space near the frogs requires for the tangent to be at least equal to 0.12. A double slip is often used in marshalling yards where the vehicle travelling speeds, and hence stresses, are very low.

Crossovers (Fig. 7.4) are used to join two parallel tracks of a given inter-axis i [3, 4]. They are implemented with two simple switches mounted with the diverging tracks in continuation and the insertion of a straight section of length l_S (called closure rail) if i > 3555 m. In this case $l_s = (i - 3555)/\text{tg}\alpha$ When two tracks need to be channelled to the right and then to the left, double or scissors crossovers take

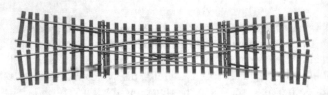

Fig. 7.3 A double slip

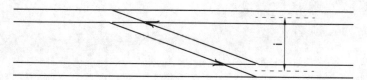

Fig. 7.4 Simplified scheme of a crossover

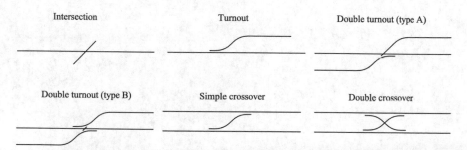

Fig. 7.5 Example of symbols used in station planes

place. This particular type of crossover is composed of four simple turnouts disposed symmetrically.

In *station planes*, where the planimetry of trucks, buildings, quays and services is reproduced (see Chap. 8), turnouts, single and double slips, crossovers and the like are represented schematically with symbols (see Fig. 7.5).

References

1. Chandra S, Agarwal MM (2007) Railway engineering. Oxford University Press
2. Profillidis VA (2022) Railway planning, management, and engineering. 5th ed. Routledge
3. Bono G, Focacci C, Lanni S (1997) Railway track (in Italian, *La sovrastruttura ferroviaria*). CIFI
4. Esveld C (2001) Modern railway track. 2nd ed. MRT-Productions

Chapter 8
Railway Lines and Stations

Abstract This chapter describes railway line configurations (i.e. single-track, double-track, triple-track and quadruple-track lines) and railway station types (e.g. way-side stations, junctions, terminals and seaport stations). After briefly introducing the main technical characteristics in terms of transportation performance, the fundamental design principles for railway lines and stations are presented, together with a few layouts.

8.1 Classification

The railway transport takes place on lines composed of one or more tracks. Every line section is generally delimited by two stations (e.g. halts, flag stations, roadside or crossing stations, junction stations and terminal stations).

Based on the track number, the lines are divided into:

- single-track lines;
- double-track lines;
- triple-track lines;
- quadruple-track lines (also known as a four-track railway);

8.2 Single-Track Lines

Single-track lines have a single track run by trains in both directions. The concurrent flows in both directions require train turnouts which take place at intermediate stations, properly equipped (Fig. 8.1).

© The Author(s), under exclusive license to Springer Nature Switzerland AG 2023   137
M. Guerrieri, *Fundamentals of Railway Design*, Springer Tracts in Civil Engineering,
https://doi.org/10.1007/978-3-031-24030-0_8

Fig. 8.1 Scheme of a single track line

8.3 Double-Track Lines

In double-track lines, each track serves only one travelling direction. In several countries, railway line directions are described as even and odd. The "even direction" is usually north- and eastbound, while the "odd direction" is south- and westbound. Trains travelling "even" and "odd" usually receive even and odd numbers and respective track and signal numbers.

In double-track lines a train overtaking occurs at intermediate stations (Fig. 8.2).

Should the flow be hindered on one of the two tracks, the operational track can be temporarily used for making trains run in both directions. It follows that, for example some trains, instead of travelling on the left track of Fig. 8.3 as in ordinary conditions, exceptionally run on the right track of Fig. 8.3 which is thus termed "illegal track"; the latter is, in fact, hardly ever equipped for being travelled in the opposite direction to the legal one (no signalling systems, often incomplete switch systems etc.). In other words, the illegal track is that one used in the reverse direction to the normal direction of travel, without any protective equipment. The illegal track is therefore used in double-track lines when the track of the normal direction of travel (i.e. the left rail) is out of service.

In order to increase safety levels, even in quite exceptional operational circumstances (e.g. emergency use), each track can be equipped for being travelled on both directions with bidirectional signalling, so as to have the same safety levels as a legal track in regular-use conditions. These double-track lines are termed bi-directional (or "banalised"), in that they have the advantage of allowing trains to run in parallel and increasing the line capacity in a travelling direction if necessary, as well as to overtake each other while running in the same direction (dynamic right-of-way).

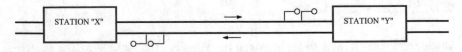

Fig. 8.2 Scheme of a double-track line

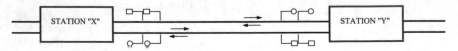

Fig. 8.3 Scheme of a bi-directional double-track line (with bidirectional signalling)

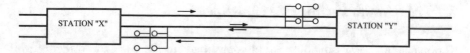

Fig. 8.4 Scheme of a three-track line with bi-directional central track

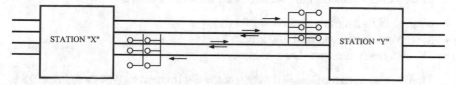

Fig. 8.5 Scheme of a quadruple-track railway (with bi-directional central tracks)

8.4 Triple-Track Lines

In triple-track lines the central track can be equipped to be travelled on both directions with bidirectional signalling, while the outer tracks keep the specialisation for their travelling directions (Fig. 8.4).

8.5 Quadruple-Track Lines

These lines guarantee higher capacities than those examined above, as well as the line specialisation for slow- and high-speed train circulation, thus avoiding promiscuity and interferences on tracks and therefore in-line delays (Fig. 8.5).

8.6 Parallel Lines and Alternative and Supplementary Routes

Parallel lines are those with a parallel and adjacent (or not too distant) horizontal alignment along a rather great length. Along the sections in parallel, in the presence of particular operating conditions (e.g. very intense traffic flows only on one line), trains can be directed from a line to another (or others) so as to distribute traffic flows.

When some traffic flow belonging to a line is directed to another line with continuity and the lines are not parallel, the line to which the flow is deviated is termed "alternative route". Should the flow be temporarily and totally deviated from a line to another (e.g. for operational interruptions due to maintenance activities), the deviated traffic stream identifies the so-called "supplementary route".

8.7   Maximum Permissible Line Speed

The maximum permissible line speed corresponds to the flank speed (defined in
Chap. 2) and must be always inferior or equal to the value of the rank speed (see
Chap. 2).

In Italy three speed regimes are provided for freight trains:

- Normal, with maximum speed of 80 km/h;
- S, with maximum speed of 100 km/h;
- SS, with maximum speed of 120 km/h.

TheUIC classifies the lines in function of the flat framework type (see Chap. 3) and
the engineering structures present. The subdivision into categories is given in Table
5.2 of Chap. 5. Trains are not allowed to circulate along a line of a given category
if their weight values per axle and per linear metre are higher than the values limit
shown in Table 5.2. In Italy the lines are classified by the infrastructure manager of
the railway network (RFI) in types A, B2, C3, D4 [1].

8.8   The Railway Stations

The railway stations are transportation infrastructures in which one or more of the
activities listed below are carried out:

- stop, crossing, right-of-way or overtaking of trains;
- passenger entry to and exit from trains;
- coach or wagon attachment to and detachment from trains;
- refuelling (for diesel trains);
- passenger assistance;
- etc.

Stations are usually sited in flat areas (to limit construction costs), wide enough
to allow the railway infrastructures to be enlarged in years to come: the smallest
stations usually cover areas of 30 m × 700 m while the biggest ones can be spread
over 500 m × 3000 m. Such areas should also be easily accessible by other transport
infrastructures (e.g. road infrastructures) and allow rails to be built in straight sections
or with very high radii of curvature, in order to lessen discomfort to travellers getting
into and out of carriages.

According to traffic flow values, stations are classified into small, medium-sized
and big; according to their position to the line, they are divided into terminal stations,
where the railway line (or one of its branches) terminates so trains must move
back before moving ahead, and intermediate stations (e.g. halts and flag stations);
according to the type of traffic, they are distinguished into travellers', freight and
mixed stations.

The Italian railway rules identify three levels according to: travellers' flow, traffic flow, size and importance of the place, business surfaces, etc. [2]:

- *1st level—large structures "Platinum"*: including big stations, high traffic levels, high quality travel service for long, medium and short routes, services for high-speed railway passengers, services for non-travelling customers, services related to the city;
- *2nd level—medium-large structures "Gold"*: including medium/large stations, high traffic levels, good quality travel service for long, medium and short routes, services for non-travelling customers;
- *3rd level—medium-small structures "Silver"*: including stations with services for long, medium and short routes, medium–low traffic levels, services for non-travelling customers; or stations for regional railway lines, generally unattended, in some cases without any passenger buildings;
- *4th level—a little crowded structures "Bronze"*: including small stations and halts with less than 200 travellers per day, often unattended, without any passenger buildings open to the public and with services for regional traffic flow.

Another interesting classification is, on the other hand, provided by the Indian Railway into block stations and non-block stations. Block stations are further classified as A class, B class and C class stations.

Also, stations can be classified under three categories:

- way-side stations: located on running lines and allowing faster train to overtake a slower train. Loop line and siding are provided. They can be of three types: halt stations, flag stations and cross stations;
- junction stations: where lines from three or more directions meet. The junction has a minimum of one main line and one branch line.
- terminal stations: where the dead end of an incoming track or at which a railway line ends or terminates.

In a railway station tracks can be subdivided into circulation tracks (running and priority) and secondary tracks (used for loading and unloading of goods and vehicle shed, manoeuvres, etc.).

For the sake of uniformity, the Italian infrastructure manager of the railway network (RFI) published guidelines [2, 3] on design standards of railway stations.

Every station must have crossing and priority track sections long enough to guarantee the transit and the yard for trains with a given length l_t expected along the railway line. Therefore, the module M_s (i.e. the minimum length) of these tracks and, consequently, of the station must meet the condition $M_s \geq l_t$. For calculating M_s the following expression is used:

$$M_s = 1.1 \cdot l_t = 1.1 \cdot (5.5 \cdot N + 2 \cdot l_{loc}) \tag{8.1}$$

where N denotes the accepted number of the in-line wheelsets (as of hour service timetable), 5.5 is the conventional interdistance between wheelsets expressed in metres, l_{loc} denotes the length of a locomotive (conventionally 25 m) and 1.1 is the multiplicative coefficient in order to guarantee a clearance safety factor. In Eq. (8.1) the module is expressed in metres.

The module of a railway line M_l corresponds to the maximum length of a train $l_{t,max}$ expected in the line. In a railway line some stations can have a lower module than that of the line they belong to ($M_s < M_l$) on condition that there are other stations which are properly distanced, equipped with crossing or priority tracks with a module at least equal to M_l ($M_s \geq M_l$) and suitable to guarantee the flow of maximum length trains $l_{t,max}$ expected along the line. In Italy the following modules are employed: 750 m (114 axes), 650, 600, 550, 500, 450, 400 m (57 axes).

8.8.1 The Way-Side Stations for Single and Double-Track Lines

Figure 8.6 represents a halt with a platform and a small travellers' shelter or cantilever roof. For single-track lines, some typical geometric layouts of way-side stations allowing faster train to overtake a slower train are given in Figs. 8.7, 8.8, 8.9, 8.10 [1, 4].

Figures 8.7, 8.8, 8.9, 8.10, 8.11, 8.12, 8.13 illustrate some station layouts with a station building (S.B.). In Figs. 8.10–8.13 platforms are connected by a subway or a foot bridge.

Fig. 8.6 Intermediate station (halt) of a single-track line (Bronze level): a shelter replaces a station building

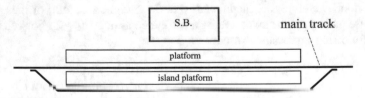

Fig. 8.7 Intermediate station (way-side station) of a single-track line, which can be used when all trains are supposed to stop

Fig. 8.8 Intermediate station (way-side station) of a single-track line, which can be used in case of frequent crossings with trains which do not stop at stations

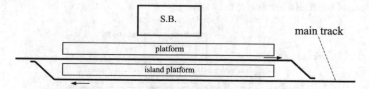

Fig. 8.9 Intermediate station (way-side station) of a single-track line with two sidings, to be used for modest traffic flow

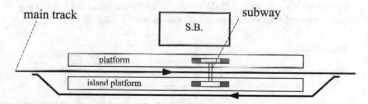

Fig. 8.10 Intermediate station (way-side station) of a single-track line, with a subway

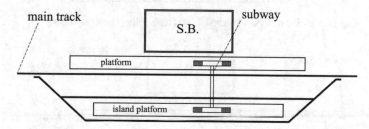

Fig. 8.11 Intermediate station (way-side station) of a single-track line, with a subway and loop lines

Figures 8.14 , 8.15 and 8.16 represent layouts of way-side stations of double-track lines. More specifically, the layout in Fig. 8.14 shows a double communication for every side by crossovers and guarantees bi-directional train movements on each track.

Figure 8.15 represents a station with an additional siding track. This layout avoids trains running the main tracks during the manoeuvres in station. The siding track is

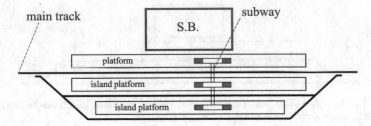

Fig. 8.12 Intermediate station (way-side station) of a single-track line, with a subway and loop lines to be used when all trains stop at the station

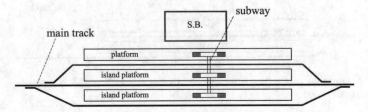

Fig. 8.13 Intermediate station (way-side station) of a single-track line with a subway, to be used in case of frequent crossings with trains which do not stop at the station

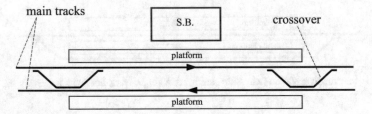

Fig. 8.14 Way-side station of a double-track line, with a siding track (layout A)

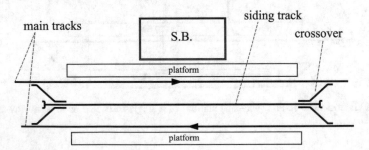

Fig. 8.15 Way-side station of a double-track line, with a siding track (layout B)

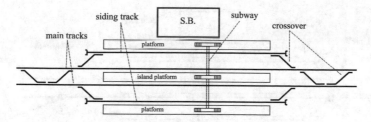

Fig. 8.16 Way-side station of a double-track line, with two siding tracks

connected to running tracks by crossovers. Figure 8.16 shows the scheme of a way-side station of double-track lines with a lateral siding track to be used when a high frequency of priority manoeuvres is expected in both travelling directions.

8.8.2 Junction Stations

They are intermediate stations in which three or more lines emerge from different directions. Such stations can be classified into [5]:

- junction stations with a single-track line and one branch line (Figs. 8.17 and 8.18);
- junction stations with a double-track line and one branch line;
- junction stations with a double-track line and two branch lines (layout A, Fig. 8.19);

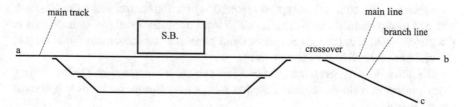

Fig. 8.17 Junction station with a single-track line and one branch line (layout A)

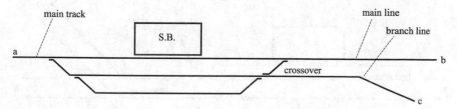

Fig. 8.18 Junction station with a single-track line and one branch line (layout B)

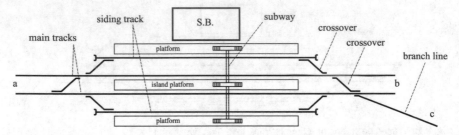

Fig. 8.19 Junction station with a double-track line and one branch line, with two siding tracks (layout A)

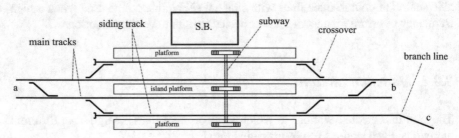

Fig. 8.20 Junction station with a double-track line and one branch line, with two siding tracks (layout B)

- junction stations with a double-track line and two branch lines (layout B, Fig. 8.20).

Figure 8.21 illustrates the simplest layout of a junction station with a double-track line and a two-branch line. Such a layout allows trains to travel from direction a to directions b and c, but it does not allow them to reverse the movement direction (i.e. the direction from b to c, or from c to b).

The point H brings greater traffic problems since, for instance, the trains coming from c interfere with those with direction a-b or b-a. Similar traffic flow problems are observed in F.

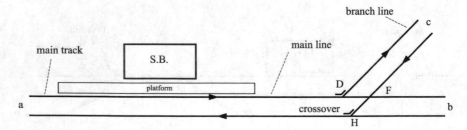

Fig. 8.21 Junction station with a double-track line and two-branch lines

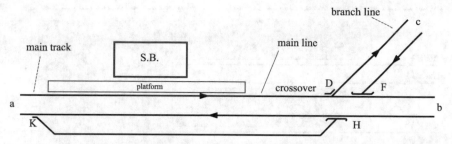

Fig. 8.22 Junction station with a double-track line and two-branch lines, plus underground track

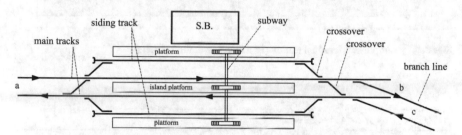

Fig. 8.23 Junction station with a double-track line, two-branch lines and two siding tracks

In order to increase the station capacity and limit interferences, an underground track (see Fig. 8.22) may be built although it entails introducing the conflict point K. Finally, the layout in Fig. 8.23 introduces siding tracks.

8.8.3 Terminal Stations

Terminal stations are where tracks, or one or more of their branches, terminate (generally with hydraulic buffers at the end of the tracks) and trains have to reverse their direction to start again. Terminals must be equipped for vehicle cleaning and maintenance and must provide users with a great many services. A basic example of a terminal station is given in Fig. 8.24 [6].

In big terminal stations (e.g. Florence, Milan, Naples, Milan etc. in Italy) there are a lot of tracks (at least 15) designed for train arrival and departure, for connection to the locomotive shed, for shelter, etc.

In several cases, however, big way-side stations (Fig. 8.25) are built where the trains arrive and depart from both the directions (e.g. Bari, Genoa, Trento etc. in Italy).

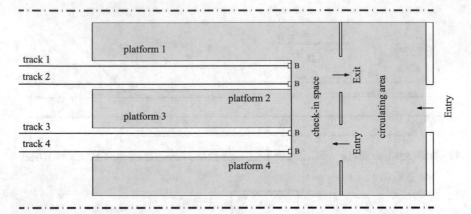

Fig. 8.24 Example of a terminal station (B: denotes hydraulic buffers)

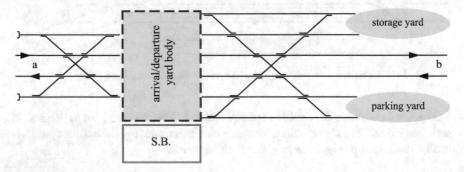

Fig. 8.25 Example of a big way-side station

8.8.4 *Yards*

A yard is the set of two or more tracks which have the aim of stabling and manoeuvring trains and are connected to one another by means of switches endowed with a shunting neck (Fig. 8.26). Yards are generally classified into: coaching yards, goods yards, marshalling yards, locomotive yards and sick line yards [6].

The shunting neck identifies the track to which trains are moved, to be then directed from one siding track to the other. Its length must be superior or equal to the longest track of the yard (usually not exceeding 800 m that corresponds to the length of the

Fig. 8.26 Example of a yard

longest trains). Should this not be the case, a number of manoeuvres are required to move, from one to another track, the rake of the wagons which are on the longest tracks: this occurs after these rakes of wagons are divided into groups of a length lower or equal to the shunting neck.

In particular cases, and only in line stations with low traffic flows, the shunting neck can coincide with the running track. However, this implies the temporary interruption of the traffic flow and low safety levels. The yards can be truncated (that is, connected to only one side with the shunting neck) or linked with both sides to the shunting neck.

In yards the tracks can be grouped in straight (Figs. 8.27 and 8.28) or in curve (Fig. 8.29). In order to optimise the spaces available, to reduce the switch (see Chap. 7) number and the track length, and to introduce the transition curves with sufficiently big radii, yards usually and properly combine both the straight and the curve track groups. By way of an example, the yard in Fig. 8.30 is obtained by developing a straight group of tracks R from the curve group, starting from the switch W. It goes without saying that further straight groups of tracks can be developed from the switches X, Y and Z. The other combination involves groups of curved tracks starting from tracks in straight groups [7].

Fig. 8.27 Yard with a straight group of tracks and aligned deviated branches

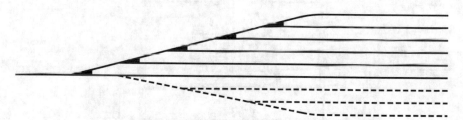

Fig. 8.28 Yard with a straight group of tracks and aligned linear branches

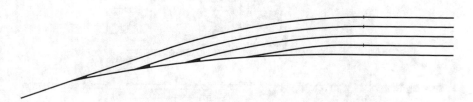

Fig. 8.29 Yard with curved tracks

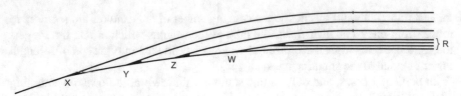

Fig. 8.30 Yard with straight and curved tracks

8.8.5 *Types of Entrances to the Platform*

The entrances to the platform can have the following configurations [3]:

- at the level of the railway tracks (Fig. 8.31);
- at a lower height than the railway tracks, when the tracks are placed on the embankment or viaduct sections (Fig. 8.32);

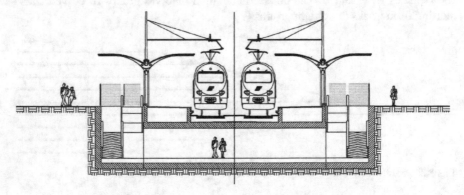

Fig. 8.31 Entrances to the platform at the level of the railway tracks [3]

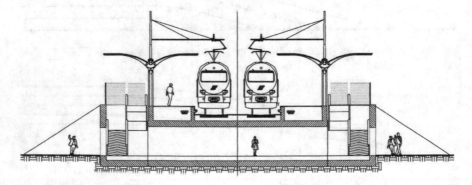

Fig. 8.32 Entrances to the platform at a lower height than the railway tracks [3]

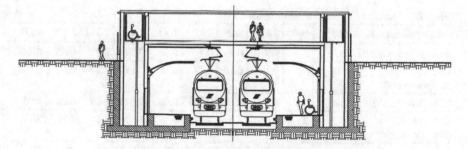

Fig. 8.33 Entrances to the platform at a higher height than the railway tracks [3]

- at a higher height than the railway tracks, when the tracks are placed in cut sections (Fig. 8.33).

8.8.6 Principles for Designing a Station Building

A station building must have a total surface S_{FV} which can be broadly estimated with the expression (8.2) or (8.3), suitable only for the stations falling within the influence areas between 50,000 and 500,000 inhabitants [8]:

- Grau's expression

$$S_{FV} = 2 K \cdot N \cdot N_v \tag{8.2}$$

- Radlbeck's expression

$$S_{FV} = 0.06 \cdot N \tag{8.3}$$

where N denotes the number of the inhabitants in the influence areas of a railway station, N_v is the number of travels made by every inhabitant in a year's time and K is a coefficient between 0.004 and 0.0017. In Eqs. (8.2) and (8.3) the surface is expressed in m^2.

In order to dimension the station facilities, the Italian infrastructure manager of the railway network (RFI) adopts the values in Table 8.1, referred to passengers expected at the peak hour. The Italian procedure is based on the following key hypotheses:

- every passenger remains 30 min in the station; therefore, the station is to be dimensioned on the basis of the total passenger number in 30 min;

Table. 8.1 Facilities surface of a building station [8]

Passengers at peak hour (n°)	100	250	500	750	1000	1500	2000	3000	4000
Hall (m²/passengers)	1	0.9	0.8	0.7	0.6	0.55	0.5	0.45	0.4
Waiting room (m²/passengers)	0.2	0.19	0.18	0.17	0.16	0.15	0.13	0.11	0.1
Information (m²/passengers)	0.1	0.095	0.09	0.085	0.08	0.075	0.07	0.06	0.05
Booking office (No./passengers)	2	3	5	7	8	10	12	16	20
Dressing room for employees (m²)	8	12	20	28	32	40	44	48	50
Luggage storage (m²/passengers)	0.1	0.095	0.09	0.085	0.08	0.075	0.07	0.06	0.05
Toilets (m²/passengers)	0.15	0.14	0.13	0.12	0.11	0.1	0.09	0.08	0.07

- during the day the traffic flow is divided into: 30% out of the total in non-peak hours (with constant values of passenger flows), 70% concentrated only in the peak hours;
- there are three peak periods a day, each lasting 90 min;
- the number of non-travelling visitors is equal to 10% of the travelling visitors.

The visitors' number P during the peak 30 min is [8]:

$$P = (1 + k_{nv}) \cdot V \cdot (0.30 \cdot \frac{t}{t^*} \cdot 0.5 + 0.70 \cdot \frac{1}{3} \cdot \frac{1}{3}) \qquad (8.4)$$

where k_{nv} denotes the percentage of non-travelling visitors with respect to travelling visitors; V is the total number of passengers at the station; t is the number of trains performing the travellers' service in an hour and t* is the number of the trains performing the travellers' service during the whole day.

Platforms have usually a height of 550 mm above the rail top surface and standard lengths of 125, 250 and 400 m in function of the traffic type (and therefore the length of trains).

Staircases for every subway and footbridge are required. Every staircase must have a width multiple of the module M (with M = 60 cm) and therefore not lower than 180 cm. Lower widths (with a mandatory minimum of 120 cm) are accepted only at stations with low passenger attendance and placed on lines unaffected by train interoperability. All stations must also accommodate staircases for the disabled with wheelchairs (i.e. a ramp for every platform and every subway and footbridge) with a width multiple of M, however not less than 180 cm. Lower widths can be accepted only at stations with low attendance and placed on lines unaffected by train interoperability.

Subways must have a width adequate for pedestrian flows, multiple of M and, in any case, not less than 3 m. The net height from the floor to the ceiling must be at least equal to 2.50 m. Cantilever roofs must be designed so as that their lengths are enough to cover the areas of entrance, waiting room, stairs and also the ramps in the presence of subways. In most cases the maximum length is not over 70 m.

For details on the evolution of the architectural styles and trends in designing stations and station buildings see references [9, 10].

8.8.7 Marshalling Yards

The marshalling yards are particular types of yards in which goods wagons in arrival are sorted out and new goods trains are formed for departure with wagons directed to the same destination (or yard). They are usually built at crucial railway junctions where more lines converge, in major industrial cities or seaports.

One or more yard bodies can be taken into consideration on the basis of the traffic flow. Each of them can serve a distinct function as follows[5]:

- train reception (arrival yard);
- wagon shunting for the next direction (direction yard);
- wagon sorting out to form new trains (sorting yard);
- train departure (departure yard);
- parking yard to add or to remove wagons;
- material storeroom (storage yard).

Based on their position, the marshalling yards can be way-side (see Figs. 8.34 and 8.35) or terminals (see Figs. 8.36 and 8.37).

In these stations marshalling operations occur by gravity in the sense that stations are given a downward slope (station "on an inclined plane"), thus producing the spontaneous movement of the vehicles which are braked and properly oriented, through the switches, along the route: arrival yard $\to$ direction and sorting yard $\to$ departure yard. The dimensioning of the track numbers of the arrival yard $N_{b.,a}$ and the departure yard $N_{b,p}$ can be performed by using the following expressions [8]:

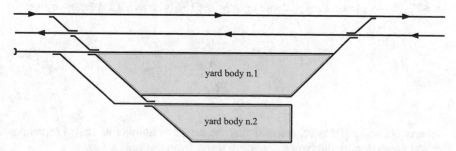

Fig. 8.34 Way-side marshalling yard with entrance to the yard body by both directions

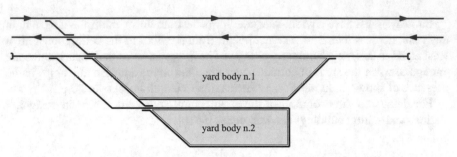

Fig. 8.35 Way-side marshalling yard with entrance to the yard body by only one direction

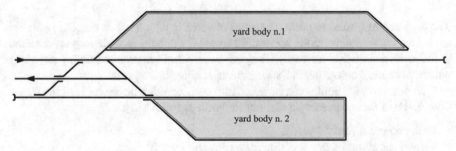

Fig. 8.36 Way-side marshalling yard with a single siding for arrival/departure trains and a single yard body

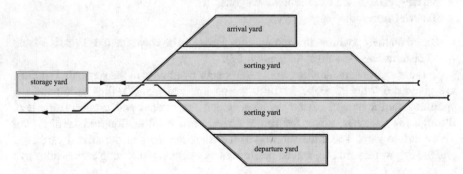

Fig. 8.37 Way-side marshalling yard with separate yard bodies for arrival and departure trains

$$N_{b,a} = \frac{N_A}{4} \tag{8.5}$$

$$N_{b,p} = \frac{N_P}{8} \tag{8.6}$$

where N_A and N_B denote, respectively, the expected number of arrival trains per day and the expected number of departure trains from the yard a day.

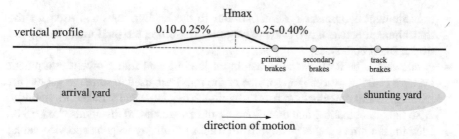

Fig. 8.38 Grades in a hump yard

In general, the marshalling yards on a horizontal plane are the most suitable for economical and safety reasons. This implies that the operations *by gravity* are carried out by means of two grades (one upward, the other downward), with the top between the arrival yard and the sorting yard (Fig. 8.38). Such a particular geometric configuration is termed hump yard and requires that the uncoupled wagons are hauled by a locomotive up to the hump vertex (of height $H_{max} = 2.5$–5 m [7, 11]). Once the hump vertex is overcome, the single wagons move on their own (as said in Chap. 1, the traction force is indeed the component of the weight force along the grade) and are directed onto the sorting yard through switches. The wagon speed is regulated by track brakes.

The locomotive shed (see Fig. 8.39) houses the repairing depots (locals for locomotive parking, cleaning and maintenance) and equipment for the complete locomotive servicing. The depots can have a circular or rectangular layout [7]. The most modern rectangular depots (Fig. 8.39) have a gantry bogie transferring the vehicles onto the yard of the shop.

In freight depots there are the following facilities [7, 11]:

- a building with the station management offices (locals for users and staff, centralised equipment management systems, etc.);
- loading equipment systems, formed of platforms aligned with tracks, 1.05 m superelevated over the top surface of the rails and with a length multiple of the wagon length (8.50 m). They can be simple or roofed;

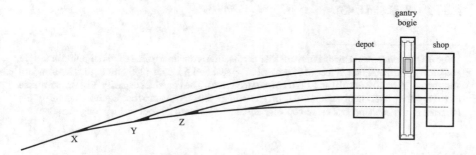

Fig. 8.39 Simplified scheme of a locomotive depot with a rectangular layout

- transhipment equipment systems, similar to the loading ones and used for the transhipment between two railway wagons. Therefore, the platforms are aligned by tracks on both sides;
- goods sheds, built at the same height as loading and transhipment equipment systems (1.05 m over the top surface of the rails) and used to store the goods (for loading on and downloading from wagons). They are usually linked to viability. Also the direct loading and downloading tracks are part of the goods shed (to be used for the direct goods passage from trucks to railway wagons and vice versa, the depot and manoeuvring tracks);
- lifting cranes, weighbridges to weigh wagons and loading gauges to verify whether the dimensions of the wagons are within the limit gauge required for railway vehicles).

8.8.8   Maritime Stations

They are railway stations built at seaports with commercial purposes [12] to allow goods to be transferred from ships to train wagons and vice versa. Goods can be directly transferred from ship to train or temporarily stored at an equipped workshop.

Loading and downloading operations for goods and containers[1] are carried out by means of sliding cranes along tracks parallel to the railway ones (with a variable gauge between 3.50 and 4.00 m), with enough operative length to serve different tracks. Alternatively, semi-mobile cranes or container gantry cranes can be used.

Figures 8.40 and 8.41 exemplify a pedestal quay crane and a sliding quay crane, respectively.

Generally, maritime stations require different yard bodies for receiving, sorting, reforming and dispatching goods trains.

Should the port spaces be inadequate for housing all those yard bodies, the sorting and reforming ones can be placed outside the port area at a reasonable distance, thus creating a further station near it [7, 11]. For direct goods loading and unloading it is necessary to arrange a landing place along the quay with two or more tracks (exceptionally one if the quay is very long). The tracks are connected to one another at a distance corresponding to the total length of the berthed ships (usually 100–150 m. see Figs. 8.42 and 8.43).

---

[1] The ISO (International Organization for Standardisation) container (often called shipping or freight container) is a metal box, 8 feet (244 cm) wide, 8 feet and 6 inches (259 cm) high. The standard lengths are 20 or 40 feet (610 and 1220 cm, respectively). The container capacity is often expressed in twenty-foot equivalent units (TEU, or sometimes *teu*). A twenty-foot equivalent unit is the measure of the containerised cargo capacity equal to a standard 20-foot (6.1 m) long container.

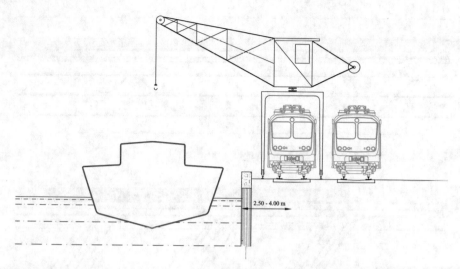

Fig. 8.40 Pedestal quay crane

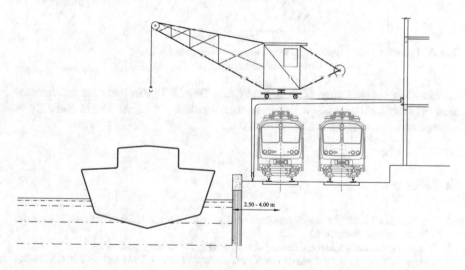

Fig. 8.41 Sliding quay crane

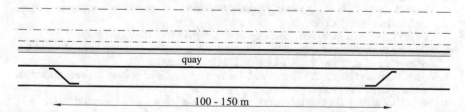

Fig. 8.42 Layout with two tracks on the quay

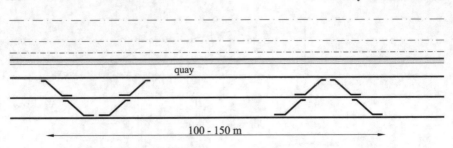

Fig. 8.43 Layout with three tracks on the quay

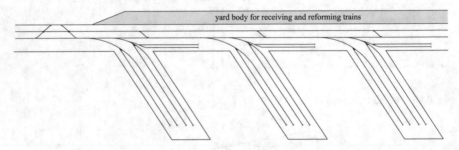

Fig. 8.44 Railway track layout for a seaport with several quays

Maritime stations have often several quays (Fig. 8.44). In this case the connection between tracks occurs with curves with a radius not below 150 m and only in exceptional cases with a radius up to 120 m.

References

1. Policicchio F (2007) Railway infrastructures (in Italian. Firenze University Press, Lineamenti di infrastrutture Ferroviarie)
2. Railway station classification. Rule, 242 Italian RFI (27/7/2007). (in Italian)
3. Design of small railway stations. Guidelines code: RFI DMO TVM LG SVI 001 A. Italian RFI. (in Italian)
4. Mayer L (2002) Railway stations (in Italian, *Impianti ferroviari*). CIFI
5. Vicuna G (1968) Railway engineering: organizations and techniques (in Italian, *Organizzazione e tecnica ferroviaria*). CIFI
6. Chandra S, Agarwal MM (2007) Railway engineering. Oxford University Press
7. Tajani F (1933) Railway transportation systems (in Italian, *Trattato moderno di materiale mobile ed esercizio delle ferrovie*). Libreria Editrice Politecnica
8. Bonara G, Focacci C (2002) Design and functional analysis of railway systems (in Italian, *Funzionalità e Progettazione degli impianti ferroviaria*). CIFI
9. Meeks CLV (1956) The railroad station; an architectural history. Yale University Press
10. Edwards B (1997) The modern station: new approaches to railway architecture. Taylor & Francis
11. Corriere F (2011) Transport infrastructures (in Italian, *Impianti ettometrici e infrastrutture puntuali per i trasporti*). Franco Angeli Editore
12. Reform of the Italian Law on Ports n. 84, 28/1/1994. (in Italian)

Chapter 9
The Railway Bridges

Abstract Railway bridges by definition are structures that carry railway lines over obstructions. These structures represent complex feats of engineering and design, and frequently involve the collaboration of a team of engineers, architects and builders. This chapter classifies and briefly describes the main types of bridges which are commonly used in ordinary and high-speed railway lines. It describes the basic concepts and assumptions for choosing a railway bridge, omitting formulas and theories.

Bridges are structures built for getting over physical obstacles including valleys, rivers, channels and gorges[1]; on the other hand, viaducts are suitable to overtop road and railway infrastructures. Despite such a distinction, the word *bridge* is often used without any consideration for the obstacle to be overcome. Conventionally bridges are structures with a span length over 5 m. When the span length, on the other hand, is within 5 m, the structures are called overpasses or culverts, depending on the obstacle type.

A bridge is composed of a superstructure constituting the horizontal span (i.e. the deck), and a substructure composed of abutments and piers, shallow and depth foundations etc. A bridge design must satisfy a series of needs due to a plethora of constraints, among which:

- hydrological and hydraulic assessments (overflow events during construction and operational phases, erosion and undermining of foundation beds, permissible vertical clearance between water level and deck etc.);
- geological and geotechnical constraints;
- geometrical constraints (e.g. with regard to the vertical alignment of the railway line);
- interferences with other infrastructures;
- useful life, construction and maintenance costs;
- etc.

---

[1] As a rule, the structure should not affect the watercourse with abutments and piers. If it were necessary to build piers in the streambed as an exception, the minimum span length between contiguous piers, measured orthogonally to the streambed, must not be lower than 40 m [1].

Fig. 9.1 A railway track above the bridge superstructure

Fig. 9.2 Railway track and deck placed below the arch of the bridge

Bridges can be classified based on:

(a) railway track level (e.g. above or below the superstructure, see Figs. 9.1 and 9.2);

(b) static (isostatic or hyperstatic) scheme and structural layout such as:

  • bridges with beams (simply supported beams, Gerber beams, continuous beams, frame bridges, etc.);
  • slab-beam bridges;
  • arch bridges;
  • cable-stayed bridges;
  • suspension bridges.

(c) material used for the structure (wood, masonry, ordinary and prestressed concrete, steel, mixed steel-concrete structure).

Generally, girder bridges and arch bridges are good solutions for short to medium spans, whereas cable-stayed bridges are utilised for medium to long spans, and suspension bridges are employed for very long spans.

For railway bridges with span lengths over 150 m the nominal useful life is always at least equal to 100 years and must be established for each project by the infrastructure manager of the railway network.

In the following sections there is a synthetic description of the most common types of bridges used in railway infrastructures. Design and calculation procedures, described in a detailed way in techno-scientific textbooks of structural engineering [2–5], do not fall within the aims of this book.

9.1 Bridges Classified According to Structural Schemes

9.1.1 Girder Bridges

In these bridges the deck is characterised from supporting beams (isostatic scheme) or continuous beams (hyperstatic scheme) which are especially subject to flexure and shear. The simply supported girder bridges are more suitable in the presence of small span lengths of the order of 40–50 m for prestressed reinforced concrete and mixed structures, and of 60–80 m for steel structures. In single span bridges the beams of the upper and lower deck edges are parallel to each other (constant construction height). This occurs independently of the horizontal and vertical alignments of the railway track. This configuration is also employed with continuous beams when the lengths of the spans are equal. In case of girder bridges, prefabrication is often extensively required (Figs. 9.3 and 9.4).

In the Gerber girder system (also known as cantilever-suspended span girder) the structure mainly consists of two components—the cantilever girder and the suspended girder. This structural system is arranged in such a way that the cantilever girder "cantilevers" over the piers, connected to the girder through a base plate, with the ends supporting the suspended spans through shear connections [2]. This structural system can be suggested when differential subsidence is expected in the

Fig. 9.3 A girder bridge with steel beams

Fig. 9.4 A continuous
concrete girder bridge

foundation [2]. The most frequent configuration has three spans with a prefabricated
central span and lateral spans cast-in-place.

Continuous girder bridges are also used to exploit the potential for prefabrication
at best. They are usually made of steel (Fig. 8.3) but a great number of them are of
prestressed reinforced concrete (Fig. 8.4). In the latter case the continuity is given
by prestressed cables placed inside the concrete beams.

9.1.2 Arch Bridges

In arch bridges the deck is supported by a semi-circular or polygonal structure mainly
subject to compression strengths and bending moments. The latter are, however,
negligible compared to those, the loads and spans being equal, on a bridge with
simply supported beams. To that end, the arch must have end-constraints in order
to provoke horizontal reactions, apart from the vertical reactions typical of a beam.
Thus, the so-called "arch behaviour" derives from the coexistence of the particular
curvilinear geometry of the structure with the horizontal reactions at the supports.

Arch bridges are more expensive than girder bridges, also for the heavy temporary
structure upon which the arch is laid during construction (centring), especially for
span lengths up to 50 m [4]. Such structures can be carried out if the foundation
soil is rocky since they convey critical horizontal actions, with the only exception
of bow-string bridges. In short, they are structures extremely suitable above all for
overtopping valleys between rocky slopes [4].

Based on the deck position, they can be classified into deck arch bridges (Fig. 9.1)
and through arch bridges (Fig. 9.2). In the latter the deck is generally suspended from
the arch by means of hangers, which are not stiff but only tension-stressed. Hangers
are carried out nearly always with vertical cables (Fig. 9.5) or with steel bars or
steel slabs to make the structure much stiffer. Rigid rods with truss configuration are
employed in some cases.

Fig. 9.5 An arch bridge
with vertical hanger
arrangements

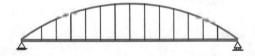

Fig. 9.6 An arch bridge with Nielsen-type hanger arrangements

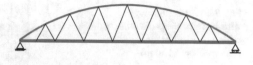

Fig. 9.7 Three-hinged wooden arch bridge

In Nielsen-type hanger arrangements (Fig. 9.6) every pair of hangers is symmetrical to a vertical axis. Further significant schemes, though rarely employed, are network arch bridges where the deck suspension is created by means of a network with a double or triple pairs of inclined hangers that cross each other at least twice. The inclined hangers with multiple intersections make the network arch bridge act like a truss, with only axial compressive and tensile forces.

With reference to the static scheme, the arch bridges are divided into [2]:

- double-fixed arch bridges: the oldest scheme, originally made of masonry but today formed of reinforced concrete and steel with parallel and vertical ribs connecting the arch structure with the deck;
- two-hinged arch bridges;
- three-hinged arch bridges: to be designed based on the pressure lines and therefore with a variable thickness structure;
- bowstring arch bridges (also called bowstring-girder bridges or tied-arch bridges): the deck transfers the vertical loads to the arch and absorbs the horizontal thrust. The deck is built in steel due the elevated tensile stresses (Figs. 9.7 and 9.8).

9.1.3 Trestle Bridges

It is a type of structure, often prefabricated, with a structural behaviour intermediate between an arch bridge and a girder bridge. Based on the span length, bridges can be designed with a single frame or aligned frames, by means of which a lower constructive height of the deck can be obtained. The most used technical solutions are:

Fig. 9.8 A trestle bridge
with inclined piers (Cadore
bridge on Piave river,
Belluno; L = 1150 m)

- small rigid-frame bridges with vertical piers, also suitable for small spans of 20–30 m like, e.g., in railway flyovers;
- rigid-frame bridges with vertical piers;
- rigid-frame bridges with inclined piers.

9.1.4   Suspension and Stayed Bridges

Suspension bridges have the deck hanging from main cables with curvilinear/parabolic configuration with the help of other vertical cables or vertical rigid elements. The main cables are fixed to the bridge ends and to the two towers.

These bridges, mostly made of steel, are employed almost exclusively for very long spans, although there are bridges with relatively modest spans (Figs. 9.9 and 9.10).

In stayed bridges the deck is supported by stays, that is, cables with an almost rectilinear configuration and anchored to the support piers. It is a steel or reinforced concrete structure, suitable to overcome medium to long spans, even higher than a thousand metres. The oblique cables can have a harp-like or fan-like configuration. In the latter arrangement the cables can be connected to a single point (saddle) or to overlapping points along the piers.

Fig. 9.9 Suspension bridge
on the Chavanon Valley,
France (span 360 m; distance
between piers 300 m)

Fig. 9.10 Stayed bridge in Piacenza (central span 192 m, lateral spans 104 m, total length 400 m)

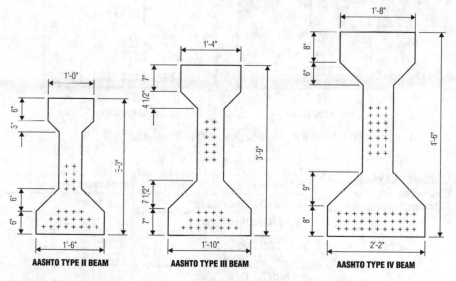

AASHTO TYPE II BEAM AASHTO TYPE III BEAM AASHTO TYPE IV BEAM

Fig. 9.11 Three different AASHTO specified prestressed concrete I-girders [7]

9.2 Bridges Classified by Material

9.2.1 Prestressed Reinforced Concrete Bridges

Although they are not traditionally employed in the railway field, in the last few years they have been extensively used in high-speed railway lines for short single-span structures of 25–50 m and precisely by adopting beams or box girders [4, 6]. Figures 9.11 and 9.12 illustrate a variety of cross-section geometries and dimensions[2] of prestressed reinforced concrete girders in function of the span values shown in Table 9.1 and suggested by the AASHTO [7].

---

[2] Expressed in the conventional American unit system.

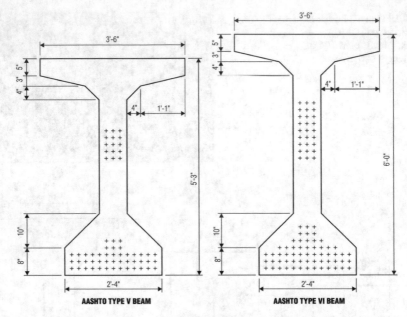

Fig. 9.12 Two AASHTO specified prestressed concrete bulb T-girders [7]

Table 9.1 AASHTO cross-section types in function of span values [7]	Cross section	Span range [m]
	AASHTO Type I girder	10.7–16.8
	AASHTO Type II girder	13.7–22.9
	AASHTO Type III girder	19.8–29.0
	AASHTO Type IV girder	25.9–36.6
	AASHTO Type V girder	33.5–44.2
	AASHTO Type VI girder	36.6–48.8

In Italy the most used deck types, especially on high-speed railway lines, are the following [6]:

- beam decks: with four prefabricated box girders or I-beams and cast slab, or two prefabricated box girders and cast-in place slab (Figs. 9.13 and 9.14). Generally, box girders are good at resisting the effects of torsion and typically do not require the introduction of bracing elements [7].
- monolithic decks: with single-cell box, two-cell box or U-beams (Figs. 9.15, 9.16, 9.17 and 9.18).

For economic reasons, on the Chinese high-speed railway (HSR) network (set to reach 70,000 km in 2035) small-sized bridges are usually built with mono- or two-cell box decks [8]. The characteristic dimensions of the decks in function of the maximum speed expected in the line (250 and 350 km/h) and the type of railway

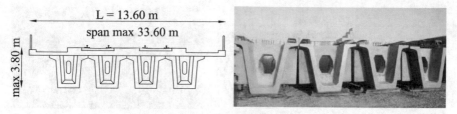

Fig. 9.13 Deck with four prefabricated box girders and cast-in place slab. *Source* Cirillo et al. [6]

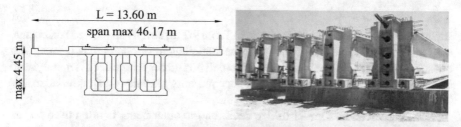

Fig. 9.14 Deck with four prefabricated I-beams and cast-in place slab. *Source* Cirillo et al. [6]

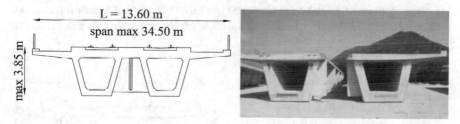

Fig. 9.15 Deck with two prefabricated box girders and cast-in place slab. *Source* Cirillo et al. [6]

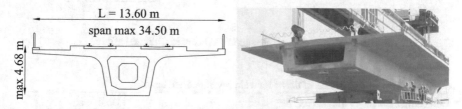

Fig. 9.16 Single-cell box deck. *Source* Cirillo et al. [6]

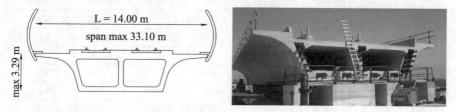

Fig. 9.17 Two-cell box deck. *Source* Cirillo et al. [6]

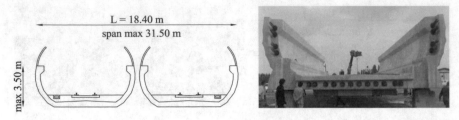

Fig. 9.18 Monolithic U-shaped deck. *Source* Cirillo et al. [6]

superstructure are given in Fig. 9.19 and in Table 9.2 for simply supported box beams, and in Fig. 9.20, Tables 9.3 and 9.4 for the static layout of continuous beams, with constant or variable beam spans. The single-cell box girder bridge deck is mainly used on lines specialised for passenger transport and only partially on freight/passenger lines.

The double-cell box girder bridge deck, on the other hand, is often used on the inter-city lines [8, 9].

Continuous box girder bridges could have either uniform or variable depth [8]. Uniform-depth continuous box girders are typically composed of two or three spans, as shown in Fig. 9.20 and in Tables 9.3 and 9.4. Material consumption is presented in Table 9.5.

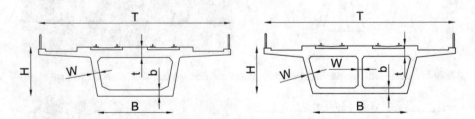

Fig. 9.19 Typical cross section of a simply supported beam in Chinese HSR [8]

Table 9.2 Typical dimensions of decks for a simply supported beam

| Geometric element | Single box | | | | Double box | |
| | 350 km/h | | 250 km/h | | 250 km/h | |
	Ballast	Ballastless	Ballast	Ballastless	Ballast	Ballastless
T [mm]	12,600	12,000	13,000	11,600	12,200	11,600
B [mm]	5500	5500	5740	5300	6500	6500
H [mm]	3072	3078	2700	2700	2700	2700
t [mm]	384	300	340	285	240	240
b [mm]	280	280	300	280	240	240
w [mm]	450	450	475	450	240	240

HSR-China, *Source* Yan et al. [8]

Fig. 9.20 Typical variable-depth single-cell bridge in Chinese HSR. *Source* Yan et al. [8]

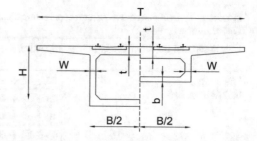

Table 9.3 Typical uniform-depth single-cell bridge in Chinese HSR

Average speed [km/h]	Span [m]	Top width T [m]	Bottom width B [m]
250 km/h	2 × 24	13.4	6.06
	3 × 24	13.4	6.06
	2 × 32	13.4	5.90
	3 × 32	13.4	5.90
	2 × 40	13.4	5.74
350 km/h	2 × 24	13.0	5.92
	3 × 24	13.0	5.92
	2 × 32	13.0	5.86
	3 × 32	13.0	5.68
	2 × 40	13.0	5.68

Source Yan et al. [8]

Table 9.4 Typical variable-depth single-cell bridge in Chinese HSR

Line speed and railway track type	Span [m]	H [m]		T [m]	B [m]	
		Mid-span of the beam	End of the side span		Mid-span of the beam	End of the side span
250 km/h-ballast track bed	32 + 48 + 32	3.40	2.80	12.2	5.56	5.74
	40 + 56 + 40	4.40	2.80	12.2	6.35	5.74
	40 + 64 + 40	5.20	2.80	12.2	6.35	5.74
	48 + 80 + 48	6.40	3.80	12.2	6.40	6.40
	60 + 100 + 60	7.20	4.60	12.2	6.40	6.40
350 km/h-ballastless track bed	40 + 56 + 40	4.35	3.05	12.0	7.70	6.70
	40 + 64 + 40	6.05	3.05	12.0	7.70	6.70
	48 + 80 + 48	6.65	3.85	12.0	7.70	6.70
	60 + 100 + 60	7.85	4.85	12.0	7.90	6.70

Source Yan et al. [8]

Table 9.5 Typical material requirement per metre on a variable-depth single-cell bridge in Chinese HSR

Line speed and railway track type	Span [m]	Concrete [m³]	Reinforcing bars [kg]
250 km/h- ballast track bed	32 + 48 + 32	11.0	1886
	40 + 56 + 40	12.6	2020
	40 + 64 + 40	13.4	2118
	48 + 80 + 48	16.0	2397
	60 + 100 + 60	18.7	3044
350 km/h- ballastless track bed	32 + 48 + 32	11.8	1796
	40 + 56 + 40	13.5	2380
	40 + 64 + 40	14.6	2422
	48 + 80 + 48	16.2	2941
	60 + 100 + 60	19.6	3115

Source Yan et al. [8]

9.2.2   Steel Bridges and Steel–Concrete Composite Bridges

The construction of steel–concrete composite decks may turn out to be useful for 40-m or higher spans. Some typical sections are shown in Fig. 9.21 [7]. In Italy steel T-shaped beams and reinforced concrete slabs cast-in place on prefabricated "predalle" slabs have been often applied to HSR lines.

Steel bridges are employed for great spans or when the deck thickness needs to be limited. The simply supported girders are often used for three or more span bridges.

Figure 9.22 shows some schemes of the early short- and medium-span railway metal bridges constructed from girders or propriety trusses (also with some structural wooden elements) adopted in the United States since the first decade of the XIX century [10].

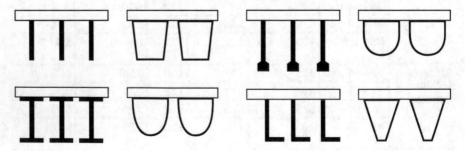

Fig. 9.21   Sections of mixed steel–concrete decks with open or closed type ribs. *Source* Tonias and Zhao [7]

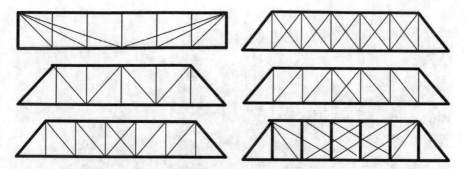

Fig. 9.22 Truss forms traditionally used in the USA. *Source* Unsworth [10]

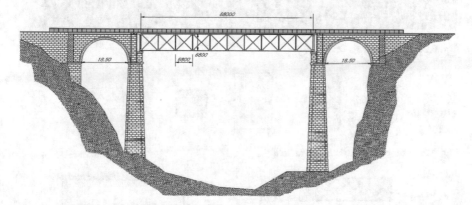

Fig. 9.23 Deck truss bridge (Aliano bridge in Civitavecchia-Orte, span 68 m)

These structures can be classified into the following main types [11]:

- deck truss bridges, rarely used, for 30–50 m spans (see Fig. 9.23);
- through truss bridges, widely used, for small spans of around 25–40 m (Fig. 9.24).

9.3 Piers

The choice of the pier cross section is correlated to functional, structural, geometrical and esthetical aspects. The most employed sections for viaducts are those shown in Fig. 9.25, while those used for watercourse crossing (bridges) are illustrated in Fig. 9.26 [10].

Fig. 9.24 Through truss bridge

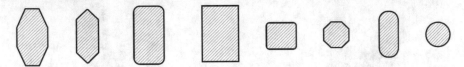

Fig. 9.25 Pier cross sections for viaducts

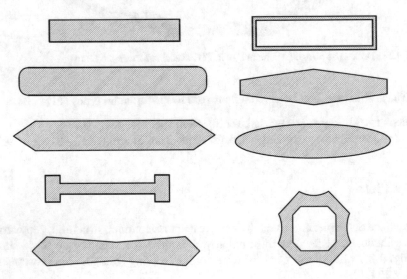

Fig. 9.26 Pier cross sections for bridges (watercourse crossing)

9.4 Actions on Railway Bridges

The actions on railway bridges will be synthetically described in the following sections. The permanent actions required for calculations are self-weights, carried permanent loads, earth pressure, hydraulic thrusts etc.

9.4.1 Deck Self-Weight

The following values can be assumed broadly [11]:

- decks with simply supported beam in prestressed reinforced concrete with spans of 30–40 m, $p = 1$ t/m^2 (9.8 kN/m^2);
- box girder bridge decks with simply supported beam (static scheme) and girders of 80–100 m, $p = 2$ t/m^2 (19.6 kN/m^2);
- decks with steel-concrete composite structure and spans of 30–40 m, $p = 0.7$ t/m^2 (7 kN/m^2);
- orthotropic steel decks with spans of 200 m, $p = 0.5$ t/m^2 (4.9 kN/m^2).

9.4.2 Permanent Loads

The permanent loads refer to the weight of both the ballast (14.7 kN/m^3) and the flat framework of the railway track.

For straight line sections, at a first approximation, the estimated permanent load is equal to around 18 kN/m^3 applied to over the entire deck width between the containment walls (paraballast), assuming a 0.80 m thickness between the rail top surface and the deck extrados.

For sections in curve the exact ballast profile must be taken into account in function of the radius; at a first approximation a load of 20 kN/m^3 can be considered. The weight of the anti-noise barriers is set in 4 kN/m^2 for a height of 4 m, measured from the deck extrados.

9.4.3 Variable Actions: Vertical Loads

The Eurocode (EN 1991-2) provides for specific load trains:

- train LM 71, to simulate the static effects produced by the normal rail traffic located on the bridge deck. LM 71 loading consists primarily of an 80 kN/m/track distributed load which is displaced over a 6.4 m length by four axles of 250 kN (Fig. 9.27);

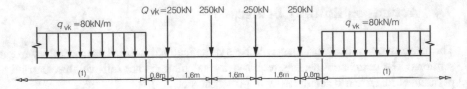

Fig. 9.27 Load Model 71 and characteristic values for vertical loads

Table 9.6 Characteristics of load configurations SW

Load model	q_{vk} [kN/m]	a [m]	c [m]
SW/0	133	15.0	5.3
SW/2	150	25.0	7.0

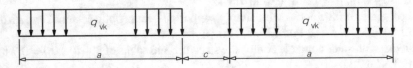

Fig. 9.28 Load Models SW/0 and SW/2

- train SW, to simulate the static effects produced by the heavy traffic flow. It is divided into two different configurations called SW/0 and SW/2 (see Table 9.6) (Fig. 9.28).

A particular load train, called *unloaded train* which is represented by a 10.0 kN/m load uniformly distributed, is used for specific verifications.

The load values examined above must be multiplied by a 1.1 coefficient for category A bridges (main railway lines) and a 0.83 coefficient for category B bridges (secondary railway lines).

The effect of lateral displacement of vertical loads shall be considered by taking the ratio of wheel loads on all axles as up to 1.25:1.00 on any one track. A point force in Load Model 71, or wheel load, may be distributed over three rail support points following the proportions 25, 50 and 25%.

On footpaths the accidental loads are schematised by a 10 kN/m^2 load uniformly distributed.

9.4.4 Dynamic Effects

The Eurocode (EN 1991-2) prescribes that: "the static stresses and deformations (and associated bridge deck acceleration) induced in a bridge are increased and decreased under the effects of moving traffic by the following:

- the rapid rate of loading due to the speed of traffic crossing the structure and the inertial response (impact) of the structure;
- the passage of successive loads with approximately uniform spacing which can excite the structure and under certain circumstances create resonance (where the frequency of excitation (or a multiple there of) matches a natural frequency of the structure (or a multiple there of), there is a possibility that the vibrations caused by successive axles running onto the structure will be excessive);
- variations in wheel loads resulting from track or vehicle imperfections (including wheel irregularities).

For determining the effects (stresses, deflections, bridge deck acceleration, etc.) of rail traffic actions the above effects shall be taken into account."

9.4.5 Lateral Actions

The key lateral actions are (i) the centrifugal force applied to a 1.80 m height from the rail top surface (running surface), calculated in function of the train load model and the maximum line speed (V = 100 km/h is only for the load model SW) and (ii) the nosing force (100 kN) acting horizontally on the top of the rails.

9.4.6 Actions Due to Traction and Braking

Traction and braking forces act at the top of the rails in the longitudinal direction of the track. Such forces are uniformly distributed along a track length L so as to exert the strongest effect on the structural element under examination. The values to consider are (Eurocode, EN 1991-2):

- for the traction force:

  - $Q_{la,k} = 33$ [kN/m] $\times$ L[m] ≤ 1.000 kN for load models LM71, SW/0, SW/2;

- for the braking force:

  - $Q_{lb,k} = 20$ [kN/m] $\times$ L[m] ≤ 6.000 kN for load models LM71, SW/0;
  - $Q_{lb,k} = 35$ [kN/m] $\times$ L[m] for load model SW/2.

In the case of a bridge carrying two or more tracks the braking forces on one track shall be considered with the traction forces on one other track.

Where two or more tracks have the same permitted direction of travel either traction on two tracks or braking on two tracks shall be taken into account.

9.4.7 Exceptional, Thermal, Wind-Induced and Indirect Actions

The bridge structures must be designed and monitored by considering also a series of additional actions: thermal, wind-induced, exceptional (e.g. catenary break, train derailment below and above the bridge), indirect actions (distortion, shrinkage and viscosity, etc.) established by the Eurocode [12]. In addition, for bridges with very long spans (strayed and suspended bridges), the aeroelastic instability phenomena must be examined in detail [3].

References

1. Design of Railway Bridges, (code: RFI DTC INC PO SP IFS 001 A). Italian RFI (in Italian)
2. Belluzzi O (1972) Solid and structural mechanics (in Italian, *Scienza delle costruzioni*), vol I. Zanichelli
3. Carotti A (2006) Solid and structural mechanics (in Italian, *Meccanica delle strutture e controllo attivo strutturale*). Springer
4. Fritz L (1979) Design of bridges (in Italian, I ponti. Dimensionamento. Tipologia costruzione). Edizioni Tecniche
5. Tonias D, Zhao J (2006) Bridge engineering. Substructure design. McGraw-Hill Professional
6. Cirillo B, Comastri P, Guida PL, Ventimiglia A (2009) High speed railway lines (in Italian, *L'alta velocità ferroviaria*). Cifi
7. Tonias D, Zhao J (2006) Bridge engineering rehabilitation, and maintenance of modern highway bridges. McGraw-Hill Professional
8. Yan B, Dai G-L, Hu N (2015) Recent development of design and construction of short span high-speed railway bridges in China. Eng Struct 100:707–717
9. Hu N, Dai G-L, Yan B, Liu K (2014) Recent development of design and construction of medium and long span high-speed railway bridges in China. Eng Struct 74:233–241
10. Unsworth JF (2010) Design of modern steel railway bridges. CRC Press
11. Policicchio F (2007) Railway infrastructures (in Italian, *Lineamenti di infrastrutture Ferroviarie*). Firenze University Press
12. EN 1991-2:2003 Eurocode 1: Actions on structures—Part 2: Traffic loads on bridges

Chapter 10
Railway Tunnels

Abstract Tunnelling is certainly one of the most fascinating but difficult civil engineering disciplines. This chapter illustrates some techniques for tunnel design starting from the rock mass classifications. For bored tunnels the New Austrian Tunnelling Method (NATM) and the ADECO-RS method are presented; while the cut and cover method is briefly analysed for artificial tunnels. It also describes some simple criteria for choosing the tunnel cross section types and the number of tubes in function of the railway line types (single or double-track lines) and length.

Tunnels are underground structures implying disturbance in a pre-existing equilibrium of the ground (a discontinuous, inhomogeneous and anisotropic material) "in conditions that are only known approximately" [1]. In function of the implemented construction system (Fig. 10.1), they are classified into:

- natural or bored tunnels, for high depths, with circular or polycentric cross-section;
- artificial tunnels, for reduced covering of ground (the depths of the top surface of the rails generally range from about 10–12 m), with rectangular or polygonal cross-section.

In general, the most stressed phase of the tunnel structure is the constructive rather than the operational stage. As a matter of fact, in natural tunnels the pre-existing actions produce overstressed conditions around the cavity of the excavation (arch effect). In the area disturbed by the excavation, with the width defining the influence radius of the face, the geomechanical characteristics of the ground decrease and cause an increase in the volume of the ground affected. Thus, the design of these underground works requires a deep knowledge of the ground in which the structure will be built, of the actions taken to excavate and the reaction of the ground to excavation (deformation response). On the basis of the nature of the ground, during tunnel advance, the face can be (see Figs. 10.2 and 10.4):

- stable, with very limited deformations (rock consistency);
- stable to a short term, with plasticization phenomena, with intrusions towards the centre of the cavity and short-term stability (this requires stabilisation interventions);

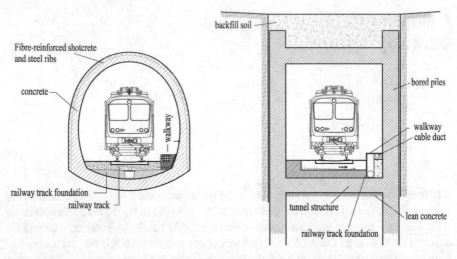

Fig. 10.1 Natural tunnel (image on the left) and artificial tunnel (image on the right)

Fig. 10.2 Stability of the excavation face: rock consistency (left), clay consistency (centre), sand consistency (right)

- instable, with no arch effect (non-cohesive or loose soils).

Tunnel design requires defining excavation stages, pre-lining, rock bolts, steel set, construction phases and times of the various components of the lining (crown arch, abutment, invert vault). The approach to design topics is interdisciplinary, thus asking for studies in the field of the applied geology, geomechanics and geotechnics/geo-engineering [2]. Some empirical models propose procedures to assist engineers in the design of stabilisation and lining works, which are based on geomechanical parameters of the ground which influence the soil behaviour. Among these are the methods proposed by Bieniawski (R.M.R System) [3] and by Barton (Q System) [4].

10.1 The Geomechanical Classification R.M.R.

The geomechanical classification R.M.R. (Rock Mass Rating), a system produced by Bieniawski [3], is based on the observation of the following five geomechanical and geostructural parameters regarding the state of the rock masses and on a correction index I_c with a value being in function of the orientation of the discontinuities and the type of the problem in question (tunnels, slopes and foundations):

1. uniaxial compressive strength of rock material ($R_1 = 1–15$);
2. rock quality designation (RQD) ($R_2 = 3–20$);
3. spacing of discontinuity ($R_3 = 5–20$);
4. conditions of discontinuities ($R_4 = 0–30$);
5. groundwater conditions ($R_5 = 0–15$).

The values of these parameters are given in Table 10.1; their sum allows the rock masses to be divided into five classes, from "Very good" (I) to "Very poor" (V):

$$\text{RMR} = (R_1 + R_2 + R_3 + R_4 + R_5) \qquad (10.1)$$

Stabilization works are undertaken for the construction of tunnel sections and are associated to each class [1, 5] Theoretically it is possible to directly select the most suitable tunnel section type to guarantee the long- and short-term stability of a tunnel by extrapolating the necessary parameters from core samples and direct measurements at the face [1]. The tunnel design information is summarized in Table 10.2.

The R.M.R. values allow the cohesion c, the internal angle of friction ϕ and the in-situ deformability of the rock mass E_d to be estimated. Some expressions obtained experimentally are:

$$c = 5 \cdot \text{RMR} \qquad (10.2)$$

$$\varphi° = \frac{\text{RMR}}{2} + 5 \qquad (10.3)$$

$$E_d = 2 \cdot \text{RMR} - 100 \qquad (10.4)$$

in which c and E_d are expressed in kPa and GPa, respectively.

Moreover, it is possible to estimate the theoretical average rate of advance (ARA_T) of an excavation of a tunnel section performed with a TBM (tunnel boring machine, see Sect. 10.5.2), expressed in m/day [3]:

$$\text{ARA}_T = 0.422 \, \text{RMR} - 11.61 \qquad (10.5)$$

Table 10.1 Classification after Bieniawski (R.M.R. index) [1]

A. Classification parameters and their ratings

Parameter			Range of values and ratings				For this low range uniaxial compressive strength is preferred		
1	Strength of intact rock material	Point-load strength index (MPa)	> 10	4–10	2–4	1–2			
		Uniaxial com-pressive strength (MPa)	> 250	100–250	50–100	25–50	5–25	1–5	< 1
		Rating	15	12	7	4	2	1	0
2	Drill core quality RQD (%)		90–100	75–90	50–75	25–50	< 25		
	Rating		20	17	13	8	5		
3	Spacing of discontinuities		> 2 m	0.6–2 m	200–600 mm	60–200 mm	< 60 mm		
	Rating		20	15	10	8	5		
4	Condition of discontinuities	Length, persistence (m)	< 1	1–3	3–10	10–20	> 20		
		Rating	6	4	2	1	0		
		Separation (mm)	None	< 0.1	0.1–1	1–5	> 5		
		Rating	6	5	4	1	0		
		Roughness	Very rough	Rough	Slightly rough	Smooth	Slicken-sided		

(continued)

Table 10.1 (continued)

A. Classification parameters and their ratings

	Rating	6	5	3	1	0
	Infilling (gouge)	None	Hard filling		Soft filling	
		–	< 5 mm	> 5 mm	< 5 mm	> 5 mm
	Rating	6	4	2	2	0
	Weathering	Unweathered	Slightly w.	Moderately w.	Highly w.	Decomposed
	Rating	6	5	3	1	0
5	Ground water	Inflow per 10 m tunnel length (L/min)	None	< 10	10–25	25–125	> 125
		p_w/σ_1	0	0–0.1	0.1–0.2	0.2–0.5	> 0.5
		General conditions	Completely dry	Damp	Wet	Dripping	Flowing
	Rating	15	10	7	4	0

p_w = joint water pressure; σ_1 = major principal stress

B. Rating adjustment for discontinuity orientations

Ratings		Very favourable	Favourable	Fair	Unfavourable	Very unfavourable
	Tunnels	0	−2	−5	−10	−12
	Foundations	0	−2	−7	−15	−25
	Slopes	0	−5	−25	−50	−60

(continued)

Table 10.1 (continued)

A. Classification parameters and their ratings

C. Rock mass classes determined from total ratings

Rating	100–81	80–61	60–41	40–21	< 20
Class No	I	II	III	IV	V
Description	Very good	Good	Fair	Poor	Very poor

D. Meaning of rock mass classes

Class No	I	II	III	IV	V
Average stand-up time	10 years for 15 m span	6 months for 8 m span	1 week for 5 m span	10 h for 2.5 m span	30 min for 1 m span
Cohesion of the rock mass (kPa)	> 400	300–400	200–300	100–200	< 100
Friction angle of the rock mass (°)	< 45	35–45	25–35	15–25	< 15

Table 10.2 RMR classification guide for excavation and support in rock tunnels [1]

Rock mass class	Excavation	Support		
		Rock bolts (20 mm diam., fully bonded)	Shotcrete	Steel sets
1. Very good rock RMR: 81–100	Full face: 3 m advance	Generally no support required except for occasional spot bolting		
2. Good rock RMR: 61–80	Full face: 1.0–1.5 m advance; Complete support 20 m from face	Locally bolts in crown, 3 m long, spaced 2.5 m with occasional wire mesh	50 mm in crown where required	None
3. Fair rock RMR: 41–60	Top heading and bench: 1.5–3 m advance in top heading; Commence support after each blast; Commence support 10 m from face	Systematic bolts 4 m long, spaced 1.5–2 m in crown and walls with wire mesh in crown	50–100 mm in crown, and 30 mm in sides	None
4. Poor rock RMR: 21–40	Top heading and bench: 1.0–1.5 m advance in top heading; Install support concurrently with excavation—10 m from face	Systematic bolts 4–5 m long, spaced 1–1,5 m in crown and walls with wire mesh	100–150 mm in crown and 100 mm in sides	Light ribs spaced 1.5 m where required
5. Very poor rock RMR < 21	Multiple drifts: 0.5–1.5 m advance in top heading; Install support concurrently with excavation; shotcrete as soon as possible after blasting	Systematic bolts 5–6 m long, spaced 1–1.5 m in crown and walls with wire mesh. Bolt invert	150–200 mm in crown, 150 mm in sides, and 50 mm on face	Medium to heavy ribs spaced 0.75 m with steel lagging and forepoling if required. Close invert

Shape: horseshoe; width: 10 m; vertical stress: below 25 MPa; excavation by drill and blast

The real (or practical) average rate of advance (ARA_R) is achieved by multiplying the ARA_T by three factors: TBM crew $F_E = 0.88$–1.15, depending on the effectiveness of the crew handling TBM and terrain; excavated length factor $F_A = 0.68$–1.20, depending on the tunnel length excavated in km and tunnel diameter factor $F_D = -0.007 \cdot D^3 + 0.1637 \cdot D^2 - 1.2859 \cdot D^2 + 4.5158$, being D the tunnel diameter in metres [3].

$$ARA_R = ARA_T \cdot F_E \cdot F_A \cdot F_D \tag{10.6}$$

10.2   The NATM Method

Since the 1960s the New Austrian Tunnelling Method—NATM has been widely used all over the world and, actually, it still is. It is based on purely observational and, in part, discretional data. The method, devised by Pacher and Rabcewicz [6], classifies the rock mass into six classes (Table 10.3) with regard to the conditions it shows when an underground opening is made. Every rock class is associated to the geomechanical design parameters, the excavation system (full or partial face), the support interventions and the time interval during which the cavity can support itself in the absence of defence works and, finally, the cross section type to take (Table 10.4). The final dimension of the cross section is chosen during the construction stage on the basis of the convergence values in the cavity measured with topographic instruments (total stations and laser rangefinders). Stresses, movements of the joints and the neutral pressures are also monitored by means of pressure cells, strain gauges and piezometers, respectively. With the NATM method, the tunnel is usually studied as a plane system, with radial stabilisation techniques only (bolts, ribs, shotcrete, concrete). By using the wedge thrust method proposed by Rabcewicz or the hyperstatic reaction method, cross sections are dimensioned by means of loads obtained from empirical or semi-empirical calculation methods (e.g. Terzaghi, Kommerel) and the help of calculation software (e.g. Stress, Kommerel etc.) [1]. The final cross section is chosen during tunnel advance by taking into consideration the actual deformation behaviour observed in situ. Therefore, such a constructive approach does not give a reliable forecasting of construction times and costs while planning the tunnel.

10.3   The ADECO-RS Method

The ADECO-RS (acronym in the Italian language for Analysis of Controlled Deformation in Rocks and Soils—Analisi delle Deformazioni Controllate nelle Rocce e nei Suoli) method was designed by Rocksoil SPA in Italy with the ambitious aim of respecting construction times and costs, independently of the excavation system, be it mechanised or conventional [1, 9].

The method is based on the in-depth analysis of the soil (carried out especially with simple compression tests, direct and triaxial cell shear tests), of the excavation actions on the advance core[1] (the advance core with a given speed V causes the stress-induced perturbation of the ground in the three dimensions) and of the deformation response of the soil to tunnel advance.

---

[1] "The volume of ground ahead of the excavation face, cylindrically shaped, and transversally and longitudinally dimensioned according to 1–1.5 times the diameter of the tunnel" [8].

Table 10.3 New Austrian tunnelling method—Rock classification after Rabcewicz-Pacher

Rock classes	I From stable to slightly brittle	II Very brittle	III Unstable to very unstable	IV Squeezing	Va Very squeezing	Vb Loose material
Characteristics	Compact material slight to medium fissuring	Heavy division into strata and fracturing, single fissures are full of clayey material-schistose intercalations	Very heavy division into strata and fracturing on several planes: fissures are full of clayey material	Very weathered rock: folded and schistose; bands of faults, well consolidated, cohesive, loose material	Completely mylonitized and weathered reduced to scree, not consolidated, slightly cohesive	Loose material, non-cohesive
Characteristics	Uni axial compressive strength σ_{gd} is greater than the tangential stress σ_τ; permanent conditions of equilibrium or guaranteed by: Measures of local protection Reinforcement of the ring of load bearing rock in the crown		The limit strength of the rock is reached and exceeded around the cross section. Supports and the creation of a ring of load bearing rock are necessary	The tangential stresses exceed the strength of the rock. The material has plastic behaviour and tends to move into the cavity reducing the cross section; intensity of the phenomenon: Medium Lateral thrusts and raising of the floor. The movements are withstood by the fully closed load bearing ring	Strong	See class Va
Influence of water	None	Unimportant	Mainly on the cavity, of the fissures	Fair	Even strong (the material tends to become soaked)	
Excavation	Full face	Full face	Top heading and bench	Division of face: I–IV	Division of face: I–VI	Division of face: I–VI

Source Lunardi [1]

Table 10.4 New Austrian tunnelling method—active stabilisation for cross-section types of tunnels with about 10 m diameter

Rock classes	I	II	III	IV	V	VI
Cross section types	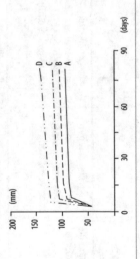					
Characteristics	Advance step 3.5/4.5 m	Advance step 2.5/3.5 m. Locally end anchored rock bolts L = 4.00 m. Steel mesh reinforcement	Advance step 1.5/2.5 m. Systematic end anchored rock bolts L = 4.00 m. Steel mesh reinforcement	Advance step 1.0/1.6 m. Systematic fully bonded rock bolts L = 4.00 m. Steel ribs Steel mesh reinforcement	Advance step 0.8/1.2 m. Systematic fully bonded rock bolts L = 4.00 m. Steel ribs Steel mesh reinforcement	Advance step 0.5/1.0 m. Systematic fully bonded rock bolts L = 4.00 m. Steel mesh reinforcement
	Lining thcknss = 0 + 0 + 30 = 30 cm	Lining thcknss = 5 + 5 + 30 = 40 cm	Lining thcknss = 10 + 10 + 35 = 55 cm	Lining thcknss = 10 + 15 + 40 = 65 cm. Tunnel invert	Lining thcknss = 10 + 20 + 45 = 75 cm, tunnel invert thcknss = 0.75 m	Thcknss = 10 + 20 + 50 = 85 cm. Tunnel Invertthcknss = 0.85 m

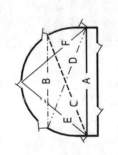

Source Lunardi [1]

Fig. 10.3 Example of tunnel excavated in partial cross-sections (NATM method)

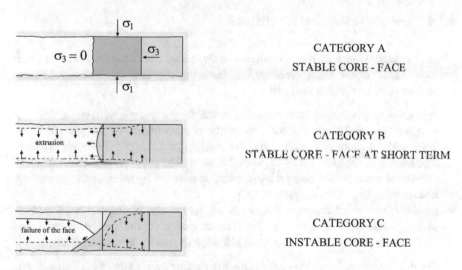

CATEGORY A

STABLE CORE - FACE

CATEGORY B

STABLE CORE - FACE AT SHORT TERM

CATEGORY C

INSTABLE CORE - FACE

Fig. 10.4   Types of the cavity behaviour [1, 8]

Unlike the New Austrian Tunnelling Method (NATM), which measures the rock mass deformation response only with the convergence (see Table 10.3), the ADECO-RS approach uses three components of the deformation response:

- advance core extrusion, measured experimentally along the longitudinal axis of the tunnel (by inserting a sliding micrometre);
- pre-convergence of the theoretical profile of the cavity[2] ahead of the face;
- convergence of the theoretical profile of the cavity behind the face.

Tunnel advance is always performed at full section (with no partial cross-sections as with NATM, see Fig. 10.3), and stabilised with specific stabilization works ahead of the face. Three distinct categories of cavity behaviour have been identified (Fig. 10.4):

---

[2] The volume of ground extruding longitudinally across the theoretical wall forming the tunnel face.

- category A or stable core-face behaviour;
- category B or stable core-face in the short term behaviour;
- category C or unstable core-face behaviour.

In the case of category A, the overall stability of the tunnel is guaranteed even in the absence of stabilisation works [1]. On the other hand, tunnels with the ground falling into category B or C require preconfinement interventions (in the short and long term) to prevent instability of the face and hence of the cavity.

For further details on the ADECO-RS method, the interested reader may refer to [1, 8].

10.4   Conservation Interventions

The conservation interventions (or, alternatively, called cavity preconfinement intervention) ahead of the face, employed in both the conventional and mechanised excavation, can be divided into [9]:

- *conservative protective* interventions, when they channel stresses around the advance core and, thus, do not change the characteristics of the ground resistance and deformability around the excavation face. Among these are drainage pipes ahead of the face, shells of ground improved by jet-grouting, shells of fibre-reinforced grout (FRG) or concrete created in advance by mechanical precutting, framework, ribs, bolts, single-cell arch;
- *advance core reinforcement* interventions, which are performed with consolidation techniques to improve the characteristics of the soil resistance and deformability before tunnelling, thus allowing the cavity to support itself.

Reinforcement interventions are frequently performed by means of, for example, (i) pre-consolidation injections by inserting fibre glass tubes (into which cement grout is injected) distributed around the perimeter of a tunnel or (ii) truncated cone "umbrellas" which are partially overlapping sub-horizontal columns of ground improved through jet-grouting, placed ahead of the face in order to generate a resistant "pre-arch" [1].

10.5   Excavation Systems

Excavation systems can depend on the excavation depth, characteristics of the ground and the operational activities. Tunnelling can be performed:

- bored for conventional tunnels:
  - with conventional method (at a full section or partial section);
  - with mechanised method (unshielded and shielded TBM);

Fig. 10.5 Face
pre-reinforcement

- cut and cover for artificial tunnels.

10.5.1   Conventional Method for Natural Tunnels

The conventional method requires the use of mechanical equipment (mechanical shovels for non-cohesive soils, drum cutters, rippers for cohesive soils, roadheaders and hammers for rocks, etc.) which are chosen on the basis of the nature of the ground, or the use of explosives employed only in case of hard rock excavations[3] [9].

The conventional method involves basic constructive phases of the same type as those concisely listed below (see Figs. 10.5, 10.6, 10.7, 10.8, 10.9, 10.10, 10.11 and 10.12) [1, 9, 10]:

- face reinforcement (layer of steel mesh reinforced shotcrete sprayed on the surface of the excavation; perforation; grout injections by means of valved fibre glass tubes; cementation);
- installation of truncated cone 'umbrellas' of drainage pipes ahead of the face;
- excavation stages;
- installation of steel ribs (IPE cross-section profiles), fibre-reinforced shotcrete or, when necessary, highperformance
    shotcrete until ribs and bolt heads are completely covered;

---

[3] The conventional excavation systems used to require nearly always advance through partial cross sections. The cavity was divided into a number of portions excavated and lined at different times, in a given sequence. The most widespread excavation methods were the "Belgian method", the "Italian method", the "Austrian-English method" and the "German method" which differed in cross-section subdivision and excavation sequence from one another.

Fig. 10.6 Installation of
drainage pipes

Fig. 10.7 Excavation

- impermeabilisation with synthetic geomembranes, 2–3 mm thick;
- invert vault installation;
- concrete casting (or installation of precast concrete segments on the walls of
  the excavation) of variable thickness in function of the geomechanical soil
  characteristics in order to complete the final lining of the tunnel cross-section.

Fig. 10.8 Steel rib installation

Fig. 10.9 Shotcrete projection

10.5.2 *Mechanised Excavation of Natural Tunnels*

The mechanised excavation is characterised by speed, continuous operation, cost and time savings and safety of workers. It is an excavation system suitable, above all, for tunnels, 3–16 m in diameter and over 2–3 km in length [11]. However, it is difficult to apply in case of small horizontal radii of curvature or logistic constraints on the transport and assembly of machines.

Fig. 10.10
Impermeabilisation

Fig. 10.11 Invert vault
installation

The machines can be divided into TBM (Tunnel Boring Machine), more exactly, in unshielded or open (cutters mainly used with hard rock) and shielded (especially employed with instable rock or under water pressure) [12]. Thanks to sophisticated automation and control systems, the shielded TBMs allow tunnel excavation and final lining with a series of precast reinforced concrete ring segments (25–30 cm thick) to be performed at the same time. After every TBM advance the tunnel cross section is complete and only needs ancillary works (railway track, cable duct, walkway, lighting systems, etc.). A TBM (Fig. 10.13) is made up of:

- the cutter head, that is a front part rotating with the cutter devices;

Fig. 10.12 Concrete casting

 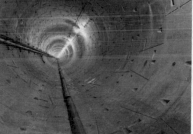

Fig. 10.13 Example of TBM and tunnel with precast reinforced concrete ring segments

- a shield consisting of a metal cylinder which closes the TBM and has a rotating head at one of its ends;
- a trailing gear composed of wagons and conveyor belts to remove excavated material and to move the precast reinforced concrete ring segments.

On the other hand, the following particular types of tunnel boring machines are appropriate to excavations in soft ground conditions containing water under pressure, loose sedimentary deposits, sand, gravel, silt, clay, formations with large boulders or high water table:

- the Earth Pressure Balance (EBP) for impermeable silty clay grounds;
- the hydroshield for permeable sandy gravel grounds.

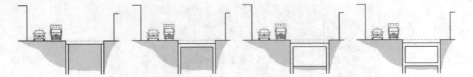

Fig. 10.14 Main construction stages of an artificial tunnel with the cut and cover method

10.5.3   Excavations of Artificial Tunnels

The construction phases of an artificial tunnel involve opening up the ground surface and excavating to the required depth, building the structure (if necessary, prefabricated) and covering the upper part of the structure with soil up to the desired level. The cut and cover method comprises the two main types of bottom-up and top-down methods.

In urban areas where the available spaces for construction sites are limited, the *cut and cover* method (also called "the Milan method") is frequently used for the construction of metros, stations, underground sections of light railways, tramways, etc.

The *cut and cover* method requires the following basic phases (Fig. 10.14):

- the area of interest is closed to the traffic;
- the excavation begins and the underground services (underground pipes, cables and equipment associated with electricity, gas, water and telecommunications) are relocated;
- support walls (bulkhead of piles) are installed for reinforcement during tunnel excavation;
- the tunnel trench is excavated;
- tunnel foundation, walls and roof are constructed using steel struts and concrete walls;
- the upper part of the tunnel structure is covered with a soil layer and the road pavement is laid;
- the area of interest is open to the public traffic.

The *cut and cover* method is more economical and commonly used at depths of 10–12 m, though it is not uneconomical even at a depth of 18 m; actually it seldom exceeds 30 m.

10.6   The Tunnel Cross Section

The loading gauge (also called gabarit) in relation to tunnels is a diagram that defines the maximum height (with respect to the top surface of the rails) and width dimensions for rolling stocks. Therefore, the loading gauge limits the size of passenger coaches,

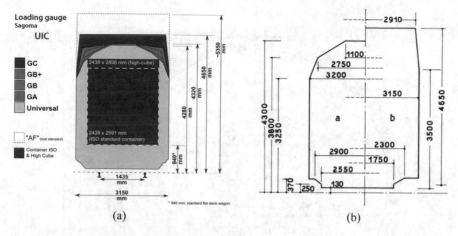

Fig. 10.15 **a** Loading gauge types according to UIC; **b** comparison between the conventional Italian loading gauge (on the left) and the "gabarit C1" sagoma (on the right)

goods wagons and shipping containers that can travel on a section of a railway track. It is evaluated in travelling conditions (kinematic gauges) (Fig. 10.15).

The minimum infrastructure gauge (MIG) is defined by given swept volume into which no obstacle must be located or intrude. This volume is determined on the basis of a reference kinematic profile and takes into account the gauge of catenary and the gauge for lower parts. Therefore, the minimum infrastructure gauge is the loading gauge of rolling stock material summed up to the minimum clearance from obstacles.

Traditionally in Italy, five different MIGs have been identified:

- MIG No. 1 for profile G1 (compatible with the European loading gauge);
- MIG No. 2 for profile B;
- MIG No. 3 for profile B plus;
- MIG No. 4 for profile C to use for renovating existing lines;
- MIG No. 5 for profile C to use on new lines.

The geometry of a tunnel cross section and the thickness of the lining are designed in conformity with the chosen MIG for a given line (or network) and on the basis of the ground type to be crossed.

The Italian infrastructure manager of the railway network (RFI) has specified seven cross sections (see Figs. 10.16, 10.17, 10.18 and 10.19) and two variants for each section (main lines and secondary lines). For double-track lines, the minimum distance between the longitudinal axes of the two tracks are given in Table 2.15, Chap. 2.

The loading gauges in the Italian railway network are distinguished into [9]:

- A (for single track) and B (double track) to use when the tunnel lining has only a protective function, or for mainly vertical thrusts on the tunnel lining;
- B1 and C, with invert vault, used for moderately thrusts on the lining of the tunnel;

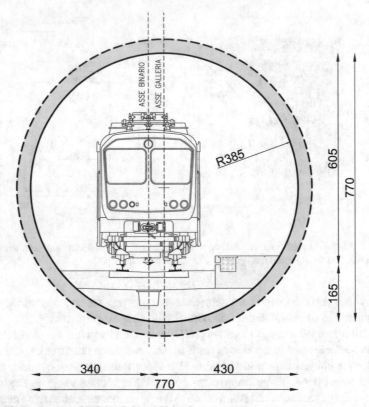

Fig. 10.16 Section type MIG No.3, V < 150 km/h

- C (for single track) and F (double track), with invert vault, used for high thrusts on the lining of the tunnel;
- D (for single track) and G (double track), with invert vault and circular section, used for high thrusts on the lining of the tunnel.

A specific design of the cross section and the lining is required only in case of asymmetrical thrusts.

The thickness of the concrete lining can be obtained through the following empirical formula [13]:

$$S = 0.083 \cdot D \tag{10.7}$$

where S denotes the thickness of the lining (in centimetres) and D the diameter of the tunnel (in centimetres).

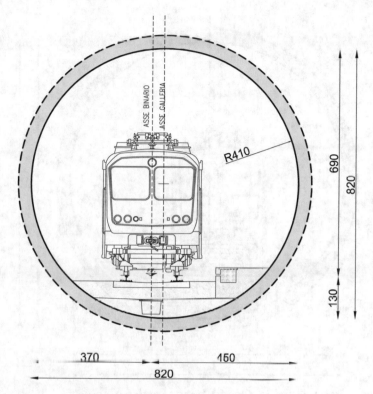

Fig. 10.17 Section type MIG No.5 (Gabarit C), V < 200 km/h

The expression (10.6) summarised the experimental observation that every 30 cm diameter requires a 2.5 cm increase in the thickness of the lining [13].

It goes without saying that the lining dimensioning in detail must be carried out with the proper methods of geotechnical and structural engineering [14, 15].

10.7 Criteria for Choosing the Number of Tubes

From the viewpoint of the horizontal alignment, it is advisable for a tunnel to develop in straight sections or in wide radius curves. As to the vertical alignment, maximum gradient limits must be respected in function of the line type (see Chaps. 1 and 2). When possible, it is also necessary to prevent the height of the rail top surface in each cross section inside the tunnel from being lower than that at the two entrances; should this be the case, a pumping system would be required to bring the water out of the tunnel.

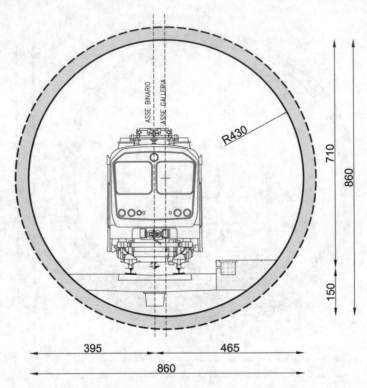

Fig. 10.18 Section type MIG No.5 (Gabarit C), V < 250 km/h

On double-track lines, a single-tube tunnel or a twin-tube tunnel can be employed. With regard to the desired safety levels, to construction costs for excavations and linings, operational costs and maximum overpressures to tunnels, the following criterion can be followed to choose the appropriate solution in function of the tunnel length L [9]:

- L ≤ 1000 m, a single-tube tunnel with aligned tracks or a twin-tube tunnel (a track for each tunnel);
- 1000 m < L ≤ 2000 m, preferably a twin-tube tunnel;
- L > 2000 m, always a twin-tube tunnel.

10.8 Construction Costs of Railway Infrastructures

The construction costs for railway infrastructures depend on numerous variables (e.g. single- or double-track lines, orography of the crossed terrain, station types and expected stations, safety systems, etc.).

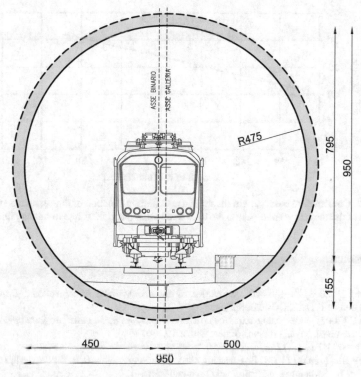

Fig. 10.19 Section type MIG No.5 (Gabarit C), V < 300 km/h

Just an indication, the construction costs per km (expressed in € million/km or M€/km) of ordinary simple-track lines are the following [16]:

- € 10 million/km for lines on flat grounds;
- € 10–30 million/km for lines on non-flat grounds;
- > € 30 million/km for mountainous routes with very long tunnel extension.

By way of an example, Fig. 10.20 shows a correlation between the construction costs per km of a single-track line and the percentage values of the length of the tunnel with respect to the total line length (total tunnel length/line length × 100).

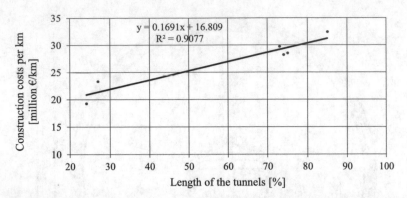

Fig. 10.20   Construction costs per km for single-track lines in function of the percentage values of the length of the tunnels with respect to the total line length (sample of Italian railway lines)

References

1. Lunardi P (2008) Design and construction of tunnels. Analysis of controlled deformation in rocks and soils (ADECO-RS). Springer
2. Barla G (2005) Underground structures (in Italian, *Sviluppi nell'analisi progettuale delle opere in sotterraneo*). Rivista italiana di geotecnica 3:11–67
3. Bieniawski ZT (1989) Engineering rock mass classifications. Wiley, New York
4. Barton NR, Lunde J (1974) Engineering classification of rock masses for the design of tunnel support. Int J Rock Mech Mining Sci Geomech 25:13–13
5. Coppolo B et al (1992) Engineering classification of rock masses using Bieniawski method (in Italian, *La classificazione geomeccanici di Bieniawski*). Geologia, tecnica e ambiente, n. 4
6. Rabcewicz L (1964) The New Austrian tunnelling method. Water Power. November 1964, December 1964, Januamnbjhry 1965
7. Dondi G, Lantieri C, Simone A, Vignali V (2014) Road construction (in Italian, *Costruzioni stradali*). Hoepli
8. Lunardi P (2014) The underground as a resource and reserve for new spaces; ADECO-RS as an effective tool to be able to realize them (part 1). In: Proceedings of the world tunnel congress 2014—tunnels for a better Life. Foz do Iguaçu, Brazil
9. Policicchio F (2007) Railway infrastructures (in Italian, *Lineamenti di infrastrutture Ferroviarie*). Firenze University Press
10. Bringiotti P (1996) Tunnelling guide (in Italian, *Guida al tunnelling*). Ediz. Pei, Parma
11. Tanziani M (2008) Mechanised tunneling (in Italian, *Scavo meccanizzato di opere sotterranee e gallerie*). Dario Flaccovio Editore
12. Maidl B, Herrenknecht M, Maidl U, Wehrmeyer G (2012) Mechanised shield tunnelling, 2nd ed. Wilhelm Ernst & Sohn
13. Chnadra S, Agarwal MM (2007) Railway engineering. Oxford University Press
14. Tesoricre G (1972) Roads, railways and airports (in Italian, *Strade ferrovie aeroporti*). UTET
15. Gattinoni P, Pizzarotti EM, Scesi L (2014) Engineering geology for underground works. Springer
16. Pyrgidis CN (2021) Railway transportation systems. CRC Press

Chapter 11
Traffic Management Systems and Railway Capacity

Abstract Traffic management systems (TMSs) help to improve railway service performance and railway network safety. TMSs enable operators to control large railway networks flexibly and proactively. This chapter presents the main systems for detecting block section occupancy, including a brief analysis of the ERTMS/ETCS system. It also describes some models for evaluating station and line capacity with automated block and mobile block systems (in homotachic regime).

Traffic management systems (TMSs) manage the following main aspects:

- headway between trains travelling on the same track;
- right-of-way regulations (on single- and double-track lines);
- crossing regulations on single-track lines;
- organisation and movement at railway stations.

Chapter 1 illustrates the relation for calculating braking distances. Figure 1.7 easily shows how incompatible such distances are with the so-called "running on sight"[1] which, for safety reasons, is exceptionally allowed in ordinary railways and, in any case, it imposes very slow speeds on vehicles.

As a matter of fact, in ordinary conditions the railway transport occurs with signalling systems (following codified rules) allowing a train to occupy a track section of a given length only when safety devices have ensured that the same section has been completely freed from any other circulating train.

This system is well-known as "block system": every track section is subdivided into successive subsections, called "block sections" (1350 m long in the Italian railway network), which allow for one train to travel at a time.

It follows that safety exigencies, along with other factors, affect the section capacity (to be meant as the maximum number of trains crossing the section in a given interval of time) and thus the line capacity.

---

[1] Safety distance, between a vehicle and the one before it, is chosen and modified, instant by instant, by train drivers autonomously, according to their perception and reaction times.

© The Author(s), under exclusive license to Springer Nature Switzerland AG 2023   201
M. Guerrieri, *Fundamentals of Railway Design*, Springer Tracts in Civil Engineering,
https://doi.org/10.1007/978-3-031-24030-0_11

11.1 Automated Systems for Detecting Block Section Occupancy

Today the main systems for detecting block section occupancy are the Axle Counter Block (ACB), the Track Circuit Block (TCB) and the Coded Current Automatic Block (CCAB).

11.1.1 Axle Counter Block (ACB)

Occupancy is controlled by means of a pair of "axle counter treadles" installed at the ends of every block section. The device counts, instant by instant, train axles at both entry and exit points. The section is considered free when the difference between the two counts (entry and exit) is equal to zero.

11.1.2 Track Circuit Block (TCB)

The system is based on an electrical circuit (track circuit) composed of a current generator, an electric link between two running rails and a rail relay detecting the presence or, more accurately, the absence of trains within a section of track. Every circuit, as long as a block section, is isolated from the adjacent ones. When the section is free, the electric circuit is run by a current which is detected by the receiver; the latter triggers the track relay, thus informing that the way is green through a connected signal.

When a train enters the block section, the first entry axle (i.e. the wheelset, see Chap. 1)—which, being built in steel, is a conductor—interrupts the electric current circulation on the track circuit (short circuit) and the relay passes to the de-excitation status, thus conveying the information that the section is occupied.

11.1.3 Coded Current Automatic Block (CCAB)

CCAB is a train protection system that uses the same technology as the track circuit block, but the current is modulated with different frequencies.

Every conventional frequency (code) is associated to a particular aspect of the trackside signal.[2] Cab-signalling systems display the trackside signals, as well as the

---

[2] A long series of trackside signals is shown on the website www.segnalifs.it.

allowable speed, location of nearby trains and dynamic information about the track ahead.

Thus, the driver is warned with optical and acoustic devices, well in advance, about the aspect of the trackside signals before they can be seen and interpreted.

In this case, the suggested safety distance (i.e. free from obstacles) is much higher than the required braking distance at a given speed.

The system also triggers an emergency braking when, after the "stop signal", the train driver does not react within 3 s [1].

In short, CCAB pursues different objectives:

- cab-signalling of the trackside signals in order to increase the safety distance;
- automated emergency braking for driver's carelessness or illness;
- safety level improvement under conditions of low visibility or risk of wrong visual perception (fog, rain, high speed, etc.).

Thus, this system is crucial, for instance, to metros where, being developed mainly or completely in tunnels, visibility conditions can be highly limited by the horizontal curves with small radii.

Considering the characteristics mentioned above, many metros are not provided with the trackside signals (being redundant), as occurs in high speed railway lines which in fact have no conventional trackside signals.

The currents are codified:

- at 4 codes (see Table 11.1) for speeds up to 180 km/h;
- at 9 codes[3] for speeds up to 250 km/h.

11.2 Automated Train Control System

11.2.1 The ERTMS/ETCS System

The ERTMS/ETCS (European Rail Traffic Management System/European Train Control System) system, or simply ERTMS, is aimed at the management, control and protection of the railway traffic and based on GSM-R (Global System for Mobile Communications—Railways) technology.

The system was worked out to ensure rail interoperability [2], especially on European high speed railway networks, in the past characterised by varied traffic and safety systems adopted by different countries.

Rail interoperability is an issue of interest for numerous objectives, such as to facilitate border crossings, open the rail signalling market, increase commercial speeds, reduce intervals between trains, decrease maintenance costs and guarantee maximum safety of the railway system.

---

[3] It also involves the codes 270* (called *average green* for allowing a maximum speed of 230 km/h) and 270** (called *super-green* for allowing a maximum speed of 250 km/h).

Table 11.1 Definition of currents at 4 codes

Code/frequency	Indication	Signal colour in	
		Line	Cab
270/4.5 Hz	Warning that at the end of the block section there are at least 2 free sections (maximum speed allowed)	Green	Green
180/3 Hz	Warning that at the end of the block section there is a free section with code 75 or 120 (slowing down required)	Green	White
75/2 Hz	Warning that at the end of the block section there is the "at danger" signal (braking manoeuvre to reach the prescribed speed until the next signal arrives)	Yellow	Yellow
120/1.25 Hz	Warning that at the end of the block section there is the entry to the deviated track (start or continuation of a stop-braking manoeuvre until the next signal arrives)	Red/Yellow	RV
–	No code (emergency braking)	Red	Red

The ERTMS integrates the functions of:

- ATP (Automatic Train Protection) which ensures the desired headway between the trains travelling on the line;
- ATC (Automatic Train Control) which triggers an emergency braking in case of a train driver's wrong behaviour.

By means of the standard ERTMS, trains receive an authorisation, that is limited in space and time, from the management system. The system provides the train driver with all the information necessary for an optimal conduct by controlling the effects of the driver's behaviour about train safe driving and by triggering the emergency braking in case of train speeds which are faster than those allowed for safety.

More specifically, every time trains can cover the segment between the point where the train receives the authorisation and the arrival point set in the same authorisation. Such an authorisation also clarifies the speeds that the train must keep along the segment under consideration (which can be also referred to several block sections) (Fig. 11.1).

There are three distinct ERTMS levels [4, 5]:

- *ERTMS Level 1:* the on-board transmission of ground information is discontinuous and supplied by fixed balises or switchable balises, placed immediately before the light trackside signals and properly connected to the signalling systems.

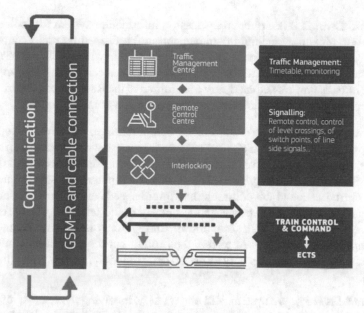

Fig. 11.1 Functional structure of the ERTMS system. *Source* [3]

The balises transmit approaching trains both the movement authorisation and unidirectional information flow necessary for driving (ground-to-train information) like, for instance, headway and speed data. Should the instantaneous speed be higher than that allowed (data shown in cabin), and the driver not brake, the automatic train braking system intervenes.

In short, this level implies the constant supervision of train movement (i.e. the onboard computer continuously supervises the maximum permitted speed and calculates the braking curve to the end of movement authority) while non-continuous communication occurs between train and trackside by Eurobalises.

The system is compatible with conventional signalling lines which can be properly equipped with ERTMS technology, thus increasing safety performances. The first-level ERTMS allows the national and interoperable traffic to coexist (Fig. 11.2).

Fig. 11.2 Working scheme of Level 1 ERTMS. *Source* [6]

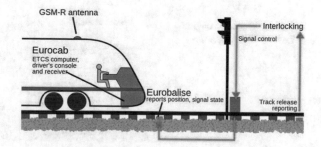

- *ERTMS Level 2*: it is a direct transmission information system for bidirectional train-trackside/trackside-train communication (performed by radio connections between a Radio Block Centre (RBC) and the train in which ATP and ATC functions are performed centrally by a traffic management centre. Position, speed and other information are automatically transmitted to the Radio Block Centre (RBC) at given intervals, thus allowing travelling to be continually monitored.

The movement authorisation is transmitted continuously to the vehicle via GSM-R together with the information on the line, like slowing down and maximum speeds.

The Eurobalises only serve as sensors for monitoring and certifying the correct train position on the line (Fig. 11.3). The on-board computer continuously processes the transferred data and the allowable maximum speeds at different sections. The system automatically intervenes when the railway traffic safety may be at risk.

On the lines equipped with second-level ERTMS only trains with second level-ETCS and GSM-R are allowed to run, thus ruling out the conventional means of transport which are able to run only by means of a ground signalling system (trackside signals).

- *ERTMS Level 3*: it is a system, still under study, involving the removal of several parts of ground equipment where there is no need for lineside signals or train detection systems on the trackside other than Eurobalises. Trains are localised by means of on-board transmitting devices which continuously communicate with the central control system. In short, train integrity is supervised by the train itself.

This level uses the mobile (or dynamic) block section (Figs. 11.3 and 11.4)

Fig. 11.3 Working scheme of Level 2 ERTMS. *Source* [6]

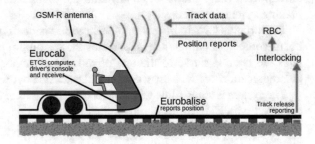

Fig. 11.4 Working scheme of Level 3 ERTMS. *Source* [6]

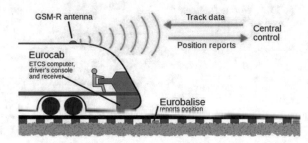

11.3 The Mobile Block Section

The system transmits directly to the cabin the maximum speed a train can keep in safety.

Every train is separated from the train ahead on the basis of the required braking distance calculated, instant by instant, in function of train speed, reciprocal distance, dynamic characteristics (acceleration and braking) and railway line alignment.

The advantage of this system for regulating traffic flows lies in the considerable increase of the line potentiality (which is inversely proportional to the block section length) especially when the flow is strongly heterotachic, as shown by numerous experiences in regulating metro traffic [7, 8].

11.4 Length Calculation of a Fixed Block Section

At the beginning of every block section there is a light signal which can change into red, yellow, green. When a train enters a block section, the corresponding protection signal becomes red, thus indicating that the section is busy (Fig. 11.5). When the train moves forward in the next section, the signal changes from red to yellow. The prospective next train is thus indicated that the section to enter is busy and therefore it must stop before this section, which will be then identified by the red colour in the light signal.

On the other hand, a signal changes into green when two or more following block sections are free.

A block section must have a length L superior or equal to the stopping distance $s(v_{max})$ of the fastest train allowed to travel along the section $(s(v_{max}) \leq L)$. By hypothesizing a uniformly decelerated motion, the stopping distance can be calculated with the relation (see Chap. 1):

$$s(v_{max}) = t_p \cdot v_{max} + \frac{v_{max}^2}{2 \cdot a} \qquad (11.1)$$

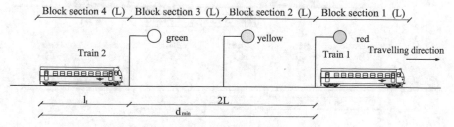

Fig. 11.5 Minimum distance between two trains on a line controlled with the automatic electric block (without a clearance distance between two successive trains)

where v_{max} denotes the speed of the fastest train (expressed in m/s), t_p is the driver's perception/reaction time (expressed in s) and a is the train deceleration (expressed in m/s²).

In Italian ordinary lines, the infrastructure manager of the railway network (RFI) assumes conventionally $V_{max} = 160$ km/h, $a = 0.82$ m/s² [9]. For a time $t_p = 3$ s, by assuming a reasonable value for the clearance between two successive trains once the breaking manoeuvre is completed, it yields $L = 1.350$ m.

When the speed V increases, the stopping distance increases as well, therefore at each speed threshold the free sections are signalled with different modes. The number of block sections necessary for stopping a train in safety is provided by the RFI by applying the following criterion:

- a block section (1350 m) for $V \leq 160$ km/h;
- two block sections (2700 m) for 160 km/h $< V \leq 200$ km/h;
- four block sections (5400 m) for 200 km/h $< V \leq 250$ km/h.

The traffic control scheme is shown in Fig. 11.6. The case of $V \leq 160$ km/h has been dealt with before, while for the other two speed regimes the following observations must be taken into consideration:

- 160 km/h $< V \leq 200$ km/h: when the train is at a point whose distance is equal to twice the block section from the beginning of a busy section, the *final green* signal is transmitted to the cabin in order to warn the train to decelerate up to the speed of 160 km/h at the signal placed at the beginning of the next block section. When the latter signal is yellow, the train must stop before entering the following

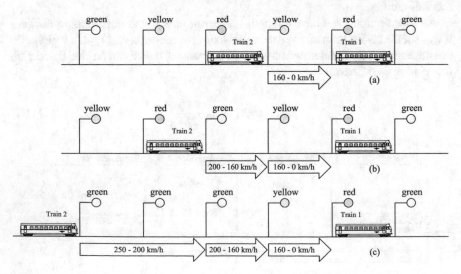

Fig. 11.6 Scheme of railway traffic control with automated block system: **a** $V \leq 160$ km/h; **b** 160 km/h $< V \leq 200$ km/h; **c** 200 km/h $< V \leq 250$ km/h

block section (i.e. within 1350 m), on the other hand, when it is green, the train can accelerate up to the initial speed;

- 200 km/h < V ≤ 250 km/h: when the train is at a point whose distance is equal to four times the block section from the beginning of a busy section, the cabin receives the signal to decelerate up to 200 km/h at the end of the two next sections and up to 160 km/h within the third next section. From this section onwards, the train stops or decelerates on the basis of the signal colour provided (yellow or green).

11.5 Line Capacity with Automated Block in a Homotachic Regime

The ideal capacity (or potentiality) C of a railway cross-section is the maximum number of trains (N) crossing such a section with a "sufficient probability of not being exceeded" in a given period of time (ΔT) generally assumed as equal to an hour, in ideal operating conditions (flow stationarity) [10].

The ideal capacity is determined in case of:

- flow homogeneity: all the trains have the same length and same performances in terms of acceleration and deceleration;
- flow stationarity: there are no variations of the flow along the section ($Q = N/\Delta T$ = const.);
- homotachic regime: all the trains run at the same speed $v_i = v$;
- negligible boundary effects: the analyses refer to sufficiently distant sections from stops or stations.

In the previous cases the minimum space distance between the ends of two trains of length l_t following each other is $d_{min} = L + L^* + l_t + f$, where f denotes the clearance between the trains once the breaking manoeuvre is completed, while L^* assumes the following values (see Figs. 11.7 and 11.8):

- $L^* = L$ for $V \le 160$ km/h (one block section, $L^* = 1350$ m);
- $L^* = 2L$ for 160 km/h $< V \le 200$ km/h (two block sections, $L^* = 2700$ m)
- $L^* = 4L$ for 200 km/h $< V \le 250$ km/h (four block sections, $L^* = 5400$ m).

The minimum time headway Δt, corresponding to the trains running with d_{min} through a cross-section, is:

$$\Delta t = \frac{L + L^* + l_t + f}{v} \tag{11.2}$$

And thus the capacity is given by

$$C = \frac{3600}{\Delta t} = \frac{3600 \cdot v}{L + L^* + l_t + f} \tag{11.3}$$

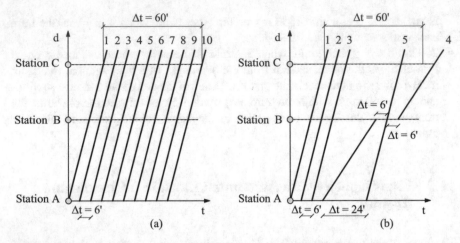

Fig. 11.7 Train trajectories in case of homotachic flow **a** and heterotachic flow **b**

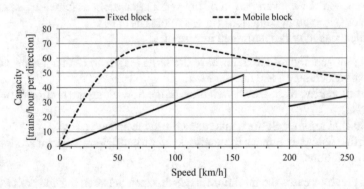

Fig. 11.8 Theoretical capacity values of railway lines with fixed block and with mobile block (l_t = 400 m, f = 200 m, t_p = 3 s, a = 0.8 m/s^2, k = 1.5)

The practical cases nearly always show a heterotachic regime. This means that a fast train must depart with a given delay compared to a slow train ahead so as to guarantee the safety distance (or that one fixed for reasons of service regularity) between the two trains at the station where the slow train, deviated to a parallel track, is overcome by the fast one.

The heterotachic regime determines a significant reduction of the line capacity compared to the homotachic regime. This is clearly evident from the train trajectories (spaces covered in function of the time, namely d = d(t)) in the examples of Fig. 11.7 (time-distance diagram). In the case (a) representing the homotachic regime, all the trains have the same speed (proportional to the gradient of the trajectory d_i(t)) and are six minutes distant from one another to guarantee a good service regularity. Therefore, in all the stations considered (A, B, C) there is a constant flow value Q =

10 trains/h per direction. In the case (b) representing the heterotachic regime, Trains 1, 2, 3 and 5 have the same speed while Train 4 is slower.

Trains 1, 2, 3, which are six minutes distant from one another, do not stop at station B. The slow Train 4 departs from A after 6' from Train 3 but it stops at station B to allow Train 5 to pass and run after 6 min after Train 4's arrival. The latter departs after further 6 min from Train 5's transition. The presence of the slow Train 4 means that Train 5 must leave A after 24 min from Train 4's departure.

In terms of flow, it is immediately clear that in 60 min 5 trains pass through A and that in the same time interval only 4 trains pass through C. Therefore, the flow in C ($Q = 4$ trains/h per direction) has more than halved compared to the homotachic regime ($Q = 10$ trains/h per direction).

11.6 Line Capacity with Mobile Block in a Homotachic Regime

For the mobile block the stopping distance is set, instant by instant, on the basis of instantaneous speed values and reciprocal distances between pairs of trains. The command/control centre transmits the position of the ahead train to every train, together with speed values and available spaces.

In varied cases already experimented the mobile block is "anchored", that is overlapped, to the conventional automated block [9]. Under this condition two trains cannot employ the same block section but, when a train overcomes a fixed yellow signal, it can avoid stopping only if the control centre warns the cabin that the required safety distance is guaranteed.

Thus, by particularising Eq. (11.3), it results:

$$C = \frac{3600 \cdot v}{L + l_t + f} \tag{11.4}$$

According to a study carried out on the German state-owned railway operator Deutsche Bahn,[4] it is necessary to multiply L by a coefficient k ($k = 1.5$) for safety reasons [11]. Taking Eq. (11.1) into consideration, the line capacity equipped with a mobile block system can be estimated with the following relation [12]:

$$C = \frac{3600 \cdot v}{k \cdot (t_p \cdot v + \frac{v^2}{2 \cdot a}) + l_t + f} \tag{11.5}$$

The speed-flow curve $C = C(v)$ obtained with expression (11.5) is known as Lehner's curve [13, 14]. By means of relations (11.3) and (11.5) it is very clear

---

[4] On these lines an "ultra-short" block is activated with the *LZB* system, which detects train position and speed and transfers discrete data ground-to-train at 25 m intervals. Thus, this system is similar to the continuous data transmission that is a characteristic of the mobile block.

that the mobile block leads to considerable capacity increases compared to the fixed block, especially at low speeds (see Fig. 11.6).

11.7   Empirical Model for Calculating the Capacities of the Lines and Stations Adopted by the RFI in Italy

The empirical relation traditionally used by the RFI to calculate the capacity C (expressed in trains/day) of a single-track line, or a double-track line with mixed traffic, with regard to a given cross-section is [15]:

$$C = k \cdot \left[N + \frac{T - t - N_v(p + i) - N_m(m + i)}{m + i} \right] \tag{11.6}$$

where

- N denotes the total number of ordinary trains included in the timetable;
- t is the no service period for maintenance activity every day [minutes];
- N_v denotes the number of passenger trains expected in the timetable;
- N_m is the number of freight trains included in the timetable;
- p is the average travelling time of passenger trains on the considered line section [minutes];
- m is the average running time of freight trains on the considered line section [minutes];
- i is the working time of the signalling devices [minutes];
- k is the reduction coefficient ranging between 0.6 and 0.8 which considers operational conditions, technical line equipment, etc.

The capacity of a station is the maximum number of trains which can arrive at a station and depart at a given time interval (generally equal to an hour).

Denoting with $C_{b,i}$ the capacity of the i-th track of the station and with C_s the total capacity of the station with N tracks, it follows [15]:

$$C_{b,i} = \frac{T}{t_{acc,i} + t_{s,i} + t'_{acc,i}} \tag{11.7}$$

$$C_s = \sum_{i=1}^{M} \frac{T}{t_{acc,i} + t_{s,i} + t'_{acc,i}} \tag{11.8}$$

In case that the occupation times at entry and exit are equal for the N tracks ($t_{acc,i} = t_{acc}$; $t_{s,i} = t_s$; $t'_{acc,i} = t'_{acc}$), the expression (11.8) becomes:

$$C_s = k_s \cdot \frac{N \cdot T}{t_{acc} + t_s + t'_{acc}} \tag{11.9}$$

where

- T is the interval time used to measure the capacity [s];
- t_s is the train dwell time [s];
- t_{acc} and t'_{acc} are the occupancy times of the analysed track at arrival and departure, respectively [s];
- k_s is the reduction coefficient for route interferences occurring to access tracks; it ranges between 0.65 and 0.8 (in case of a station under non-saturated conditions).

The time interval t_s corresponds to the time interval between the instant when the train stops at the station in order to allow passengers to get off at the arrival and the departure instant after new passengers have got on. An estimation of t_s can be made by using the following expressions [15]:

- for local trains:

$$t_s = 2 + 0.7604\,D + 0.7627\,S + 3.526 \qquad (11.10)$$

- for express trains:

$$t_s = 2 + 1.1535\,D + 1.2541\,S + 2.7873 \qquad (11.11)$$

in which t_s is expressed in s, D is the number of the passengers getting off the train and S is the number of the passengers getting on.

11.8 Capacity Indices

If the capacity C and the expected traffic demand $Q = Q(x, t)$ at the cross-section of interest x and at time t are known, further traffic variables can be calculated for characterising the operational conditions of a railway line:

- the degree of saturation $\rho(x, y)$:

$$\rho(x,t) = \frac{Q(x,t)}{C} \qquad (11.12)$$

- the reserve capacity $R(x, t)$

$$R(x, t) = C - Q(x, t) \qquad (11.13)$$

- the percentage capacity rate

$$R_p(x, t) = \frac{R(x, t)}{C} \cdot 100 = \frac{C - Q(x, t)}{C} \cdot 100 \qquad (11.14)$$

In the planning and design stages, the flows $Q(x, t)$ are estimated with the quantitative models which are typical of the traffic and transportation engineering [9, 16].

Traffic flows, line capacities[5] and the traffic variables described above are used in different design phases and are basic information to carry out financial and economic analyses [17] required in feasibility studies of railway infrastructures [18]. Such analyses, supported by the corresponding financial and economic indicators (i.e. Economic Net Present Value—ENPV,[6] Financial Net Present Value—FNPV, Economic Internal Rate of Return—EIRR,[7] Benefit–Cost Ratio—BCR and Payback Period) are expressly required by the European Commission [19] which considers them as valuable methods for identifying the co-financeable "major projects" of transportation infrastructures [19, 20].

References

1. Trenitalia (2008) STB: http://www.fastmobilita.it/docimmagini/s2f0_1.pdf
2. Directive 2008/57/CE on the interoperability of the rail system within the Community
3. European commission, Mobility and Transport, ERTMS. https://transport.ec.europa.eu/transp ort-modes/rail/ertms/how-does-it-work_en. Accessed on 20 Sept. 2022
4. Senesi F, Marzilli E (2008) ETCS European train control system. CIFI
5. Policicchio F (2007) Railway infrastructures (in Italian, *Lineamenti di infrastrutture Ferroviarie*). Firenze University Press
6. European commission, Mobility and Transport, ERTMS. https://transport.ec.europa.eu/transp ort-modes/rail/ertms/how-does-it-work/etcs-levels-and-modes_en. Accessed on 21 Sept. 2022
7. Ning B et al (2010) Advanced train control systems. WIT Press
8. Wang Y, Ning B, De Schutter TBB (2016) Optimal trajectory planning and train scheduling for urban rail transit systems. Springer
9. Ferrari P (2001) Planning of transportation systems (in Italian, *La pianificazione dei trasporti*). Pitagora Editrice Bologna
10. UIC Code 406, Capacity. 1st edition, June 2004
11. Sporleder H (1989) Continues automatic train control and cab signalling with the LZB 80. In: International conference on main line railway electrification, York, UK

---

[5] The capacity in terms of transported passengers (passengers/hour per direction) is calculated by multiplying the capacity C by the number of passengers who can be transferred by every train.

[6] The ENPV is calculated as follows: $ENPV = \sum_t B_t(1 + i_t)^{-t} - \sum_t C_t(1 + i_t)^{-t}$

where B_t and C_t are, respectively, benefits and costs occurring at time t and it is the social discount rate. The main cost items considered: infrastructure construction costs, vehicle purchasing costs, infrastructure and vehicle operation and maintenance costs, energy costs, accident costs, climate change costs, user travel costs. The ENPV is estimated by taking into consideration a railway service life conventionally set in 30 years [18, 19].

The project is assessed positively when the ENPV is positive and vice versa [17]. Among alternative project scenarios the most suitable is generally that corresponding to the highest ENPV.

[7] The EIRR is the discount rate value which yields $ENPV = 0$. Generally, the authorisation of a transportation project is granted if the following conditions are satisfied: $ENPV > 0$, Ratio Benefits/Costs > 1 and EIRR > social discount rate [17]. Between the two scenarios the most suitable is that one with higher EIRR [9].

12. Delfino A, Galaverna M (2003) Mobile and fixed block: carrying capacity analysis. Ingegneria Ferroviaria 6:555–565
13. Papacostas CS, Prevedouro PD (2001) Transportation engineering and planning, 3rd ed. Prentice Hall
14. Vuchic VR (2007) Urban transit systems and technology. Wiley
15. Bonara G, Focacci C (2002) Functionality and design of railway infrastructures (in Italian, *Funzionalità e Progettazione degli impianti ferroviaria*). Cifi
16. de Dios Ortúzar J, Willumsen LG (2001) Modelling transport, 4th ed. Hoepli
17. Corriere F, Di Vincenzo D, Guerrieri M (2013) A logic fuzzy model for evaluation of the railway station's practice capacity in safety operating conditions. Arch Civ Eng 59(1):3–19
18. Senn L, Ravasi M (2001) Investing in infrastructures (in Italian, *Investire in infrastrutture*). Egea
19. Guide to Cost-Benefit Analysis of Investment Projects. Economic appraisal tool for Cohesion Policy 2014–2020. European Commission, 2014
20. Guerrieri M (2022) Hyperloop, HeliRail, Transrapid and high-speed rail systems. Technical characteristics and cost-benefit analyses. Res Transp Bus Manage 43
21. Guerrieri M (2022) Catenary-free tramway systems: functional and cost-benefit analysis for a metropolitan area. Urban Rail Transit 5(4):289–309

Chapter 12
High-Speed Railways, Maglev and Hyperloop Systems

Abstract In the last decades, several innovative technologies have been developed with the purpose to produce increasingly fast and efficient transportation systems, including high-speed railways, the transrapid and the hyperloop systems. This chapter briefly describes such systems and summarized some of their technical characteristics as well as the related construction costs.

12.1 High-Speed Railways

There is not a single definition of a high speed railway (HSR) system. A speed of 200 km/h was initially set as a threshold limit of distinction between conventional railway and high-speed railway systems [1]. The International Union of Railways (UIC) considers a commercial speed of 250 km/h as the lower speed limit for the identification of HSR. High-speed railway systems typically serve longer routes than 400 – 500 km with few intermediate stops.

In HSR the two electrification systems 15 kV/16.7 Hz and 25 kV/50 Hz AC are generally adopted.

HSR lines, compared to ordinary ones, are characterized primarily by the increase in aerodynamic resistance to which trains are subject (see Chap. 1), the greatest dynamic loads (see Chap. 5) and the remarkable environmental acoustic and vibrational impact. The alignment design criteria follow the rules specified in Chap. 2. Some types of HSR cross-sections on the viaduct and gallery sections are represented and described in Chaps. 9 and 10, respectively.

For speed V values below 300 km/h, the area A of single-tube tunnels can be estimated approximately with the relationship (12.1), obtained with data taken from [1, 2], in which A and V are expressed in m^2 and km/h, respectively:

$$A = 0.428 \cdot V - 28.571 \qquad (12.1)$$

For instance, for V = 300 km/h the area is: $A \approx 100 \ m^2$.

In Italy, for comfort reasons, the average deceleration of high-speed trains (e.g. ETR 600 "Freccia Argento", ETR 500 "Freccia Rossa) is set at 0.75 m/s^2 [3] with

© The Author(s), under exclusive license to Springer Nature Switzerland AG 2023   217
M. Guerrieri, *Fundamentals of Railway Design*, Springer Tracts in Civil Engineering,
https://doi.org/10.1007/978-3-031-24030-0_12

consequent high values of breaking distances. For example, starting from an initial speed V = 300 km/h, a braking distance of 4.6 km is required.

For these reasons, HSRs need specific automated train control systems for operating traffic in safe conditions (e.g. the ERTMS/ETCS 2nd Level System is mandatory in Italy).

Table 12.1 summarises the main technical characteristics of the high-speed railway lines according to the most recent Italian design guidelines [3].

Table 12.2 gives a list of the high-speed railway network extensions in operation and under construction in several countries.

Figure 12.1 shows the types of grade separation in HSR and conventional railway lines, from exclusive exploitation to fully mixed.

Exclusive exploitation (EE) requires the construction of new high-speed infrastructures and high-speed trains must use only high-speed tracks. EE is characterized by a complete separation between high-speed and conventional services, each one with its own infrastructure. In fully mixed mode (FMM) both conventional and high-speed trains can run on lines of conventional and high-speed tracks; therefore, FMM allows for the maximum flexibility of the railway network.

12.1.1   Construction Costs of High-Speed Railways

The construction cost of a high-speed railway line is very variable and depends on the country and context. In [4] 166 HSR projects around the world were analysed; it was found that HSR construction costs range from € 6 million/km to € 45 million/km with an average value of € 17.5 million/km. Figure 12.2 shows a typical percentage incidence of HSR construction costs.

12.2   Maglev Systems

Maglev (abbreviation of "magnetic levitation") is a transportation system that suspends, guides and propels trains using magnetic levitation. The most popular Maglev systems are Transrapid, Linimo and ECOBEE. The term levitation indicates that a suspended magnet is free to continuously move at a certain distance from the fixed track. The principle of electromagnetic levitation is schematized in Fig. 12.3 [8]. The flux φ of the magnetic circuit is defined as:

$$\varphi = \frac{F_m}{R} \tag{12.2}$$

In which F_m is the magnetomotive force and R is the reluctance, the value of which is

Table 12.1 Main technical characteristics of high-speed railways (Italian Standard, *Source* [3])

Maximum speed	V = 300 km/h
Gauge	1435 mm
Minimum planimetric radius	$R_{min} = 550$ m for V = 300 km/h
Recommended planimetric radius	> 7000 m
Superelevation (maximum value)	$h_{max} = 105$ mm
Uncompensated accelerations	0.6 m/s^2
Transverse jerk	0.15 m/s^3
Transition curve types	Cubic parabola or clothoid
Minimum length of transition curves	330 m
Minimum length of straights	170 m
Distance between the two tracks	5 m
Maximum platform width	12.5 m
Maximum slope	18‰
Minimum vertical radius	$R_{min} = 0.35 \ V_{max}^2$
Minimum tunnel area	82 m^2
Loading gauge	Gabarit C
Rail type (LWR – long welded rail)	60 UNI
Spacing of sleepers	60 cm
Rail fastenings	Pandrol clip
Ballast (minimum depth)	35 cm
Sub ballast (minimum depth)	12 cm (bituminous concrete)
Maximum load per axle	25 t
Electrification system	2 × 25 kV, c.a. – 50 Hz
Turnout (geometric characteristics)	tg 0.022 (V = 160 km/h) and tg 0.015 (V = 220 km/h)
Signalling and control component	ERTMS/ETCS 2nd Level
Switching distance	Every 24 km (to switch from one line track to another)

$$R = \frac{l}{\mu A} \qquad (12.3)$$

where l, A and μ are the length, the area of flux path and the magnetic permeability, respectively. The permeability of free space is $\mu_0 = 4 \ \pi \ 10^{-7}$ H/m. For steels, the permeability is in the range 2000–6000 or higher than μ_0.

Table 12.2 High-speed railway network extensions in several countries

Country	In operation [km]	Under construction [km]	Total network [km]
China	31,043	7207	38,250
Spain	2852	904	3756
Japan	3041	402	3443
France	2814	0	2814
Germany	1571	147	1718
Turkey	594	1153	1747
Italy	896	53	949
South Korea	887	0	887
USA	735	192	927
Saudi Arabia	453	0	453
Iran	0	787	787
Taiwan	354	0	354
United Kingdom	113	230	343
Belgium	209	0	209
Morocco	200	0	200
Switzerland	144	15	159
The Netherlands	90	0	90
Denmark	0	56	56
Total	45,996	11,146	57,142

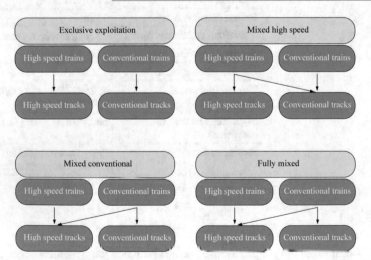

Fig. 12.1 HSR grade separation (adapted from [4])

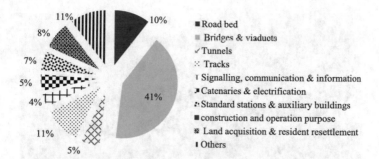

Fig. 12.2 Typical percentage incidence of HSR construction costs (from data by [5–7])

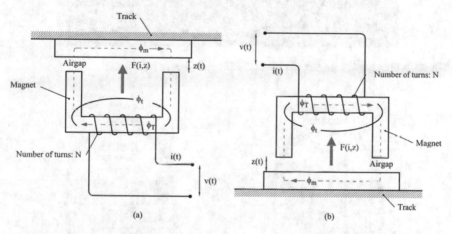

Fig. 12.3 Principle of electromagnetic levitation: **a** Electromagnet-track configuration; **b** Suspension of an object by an electromagnet fixed to the ground (adapted from [8])

Finally, the magnetomotive force F (expressed in ampere-turns) is equal to the effective current flow applied to the magnetic core made from ferromagnetic materials.

$$F = N \cdot i \tag{12.4}$$

Two predominantly distinguished types of maglev (Fig. 12.4) technology [10, 11] can be used: Electromagnetic Suspension System (EMS) and Electrodynamic Suspension System (EDS) (Fig. 12.5). Essentially, the EMS controls (by electronic devices) the electromagnets which are placed in the vehicle and attract it to the magnetically conductive guideway (usually a steel track). On the other hand, EDS uses the same polarity of magnets to levitate trains by repulsive force from the induced currents located in the conductive guide ways and keep trains and guides separate. Some of the main characteristics of EMS and EDS are given in Table 12.3 [12].

Fig. 12.4 Example of a maglev train and infrastructure (*Source* [9])

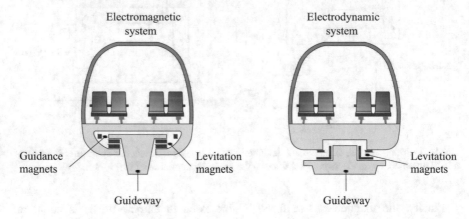

Fig. 12.5 Layouts of maglev technology types

The lift magnets are usually placed in the vehicles, propelled by linear motors, and the ferromagnetic track is mounted underneath the guideways. The trains of maglev systems have no wheels, transmission and axles, and use non-contact magnetic levitation, guidance and propulsion systems.

Generally, two types of linear motors are used for contactless propulsion:

- LIM (Linear Induction Motor) is used mainly for low-speed propulsion;
- LSM (Linear Synchronous Motor) can be used for both low- and high-speed propulsion.

The Transrapid is the first maglev-based railway system in the world and can be considered the most advanced electromagnetic system able to reach speeds superior

Table 12.3 Main characteristics of EMS and EDS systems

Item	EMS	EDS
Type of mode	Attraction mode	Repulsive mode
Magnets	Iron cored electromagnets	Superconducting coils
Guideway	10–15 mm	100–150 mm
Guideway components	Laminated strips	Aluminium strips
Stability	Inherently unstable	Dynamically stable
Feedback control	Necessary to maintain dynamic stability	Necessary
Compatible drive system	Linear induction motor	Linear synchronous motor
Example	Transrapid	MLX

Source [12]

to 500 km/h. The speed can be adjusted depending on the frequency of the alternating current supplied. Braking and stopping of the train are performed by generating a magnetic force in the opposite direction. The Transrapid uses electromagnets for levitation and LSM for train propulsion. The clearance between the lift magnet and the guideway is 10 mm.

The Transrapid maglev train runs over a double-track guideway. The trains are fully automated (self-driving).

The suggested minimum planimetric radii in function of the design speeds and for a transversal inclination of the track equal to 12 degrees are: 3444 m for 400 km/h, 4360 m for 450 km/h and 5382 m for 500 km/h. The maximum track slope may reach 10%. The switches are composed of continuous steel box beams with lengths between 78 and 148 m. Tunnels in the Transrapid system can be of two types: single-tube double-track or double-tube single-track tunnel. For single-tube double-track, the diameter is 8.20 m [13]. The main characteristics of the latest Transrapid 09 are given in Table 12.4.

12.2.1 Construction Costs of Maglev Lines

The construction costs vary from project to project. A summary of the costs and lengths for different maglev projects are presented in Table 12.5: it can be noted that the costs range from € 19 million/km to € 168 million/km with an average value of € 88 million /km.

Table 12.4 Transrapid
09—main technical
specifications (data from [8,
14])

Item	Value
Number of sections	3
Total length	75.8 m
Vehicle width	3.7 m
Vehicle height	4.25 or 3.35 m (from guideway gradient)
Inner width of carriage body	3.43 m
Inner height of carriage body	2.1 m 2.05 m (entrance door area)
Empty weight	169.6 tons
Full loaded weight	210 tons
Design speed	505 km/h
Train carrying capacity	449 persons
Design pressure	± 5500 Pa
Sealing time constant	$\tau > 20$ s

Table 12.5 Maglev project costs

Maglev project (Year of Cost)	Country	Total project cost including trains (Mln €)	Length (km)	Project cost per km (Mln € /km)
Shanghai (2004)	China	1827	29.93	61.02
Linimo (2005)	Japan	1031	8.85	116.48
Incheon Airport (2012)	SK	325	6.12	53.19
Changsha (2016)	China	609	18.51	32.90
Shinkansen (2016)	Japan	81,308	500.02	162.61
Baltimore Washington DC (2016)	USA	10,841	64.37	168.41
U.S. Maglev Network (2013)	USA	902,841	46.67	19.34

Source [7, 15]

12.3 The Hyperloop System

The hyperloop (HL) system is a relatively new transport technology at the theoretical and experimental stages of development. It is claimed to offer greater performances compared to conventional HSR (High-Speed Rail) and APT (Air Passenger Transport) systems, mainly regarding speed, transport costs, energy consumption and safety. The Hyperloop (Fig. 12.6) is based on the idea of vehicles running inside an almost-vacuum environment [16]. To date, the scaled test runs in absence of human passengers and trains have been conducted to a maximum speed of 457 km/h. The Hyperloop was designed to overcome the problem of train resistance force (see

Fig. 12.6 Rendering of the hyperloop system (*Source* [18])

Chap. 1) at high speeds by operating in a low-pressure environment i.e. in a uniform thickness steel, partially evacuated, cylindrical tube. The lower air density (and low pressure of around 0.015 psi) obtained by pumps allows the resistance force to be limited at high speeds. The speed of the capsule (up to 1220 km/h) can be uninterrupt-edly controlled and modified by varying the frequency of the alternating current of the propulsion system (Linear Induction Motor). According to [16, 17], the hyperloop system is assumed to consist of five main components:

- the line/tube including at least two parallel tubes and the stations along them, which enable operations of the HL vehicles in both directions without interfering with each other and embarking and disembarking of passengers, respectively;
- the fleet of HL vehicles can consist of a single and/or a few coupled capsules (carrying capacity P = 28 people each) moving at high speed. Each capsule is 1.35 m wide, 1.1 m high and 30 m long;
- the capsules should be separated within the tube by approximately 37 km on average in order to guarantee safety during operation. The system requires an Automatic Traffic Control (ATC) system similar to an ETCS level 3 moving block system;
- the vacuum pumps running continuously at various locations along the length of the tube to maintain the required pressure, despite any possible leaks through the joints and stations;
- the vehicle control system operating along the line(s)/tube(s);
- the maintenance systems for all previous components.

The main characteristics of the proposed hyperloop system are shown in Table 12.6.

Taking into account the maximum speed of the system (V = 1220 km/h) and a maximum deceleration a = 1.5 m/s^2, under the hypothesis of a uniform deceleration motion, it results in a braking distance s = $(V/3.6)^2/(2a)$ = 38.28 km. In steady state traffic conditions [19], it results: vehicle density k = 1/s = 1/38.28 = 0.026 veh/km,

Table 12.6 Main characteristics of the hyperloop system (adapted from [16])

Maximum speed (estimated)	V = 1220 km/h
Propulsion	Linear motor
Partially evacuated tube material	Steel
Partially evacuated tube thickness	20–23 mm
Partially evacuated tube supports	piers
Partially evacuated tube pressure	0.015 psi (100 Pa)
Partially evacuated tube diameter	2.23 m
Spacing of the piers	20–40 m
Transition curve types	Clothoid or sinusoidal curve
Capsule width	1.35 m
Capsule maximum height	1.10 m
Capsule weight	3100 kg
Capsule cost	$\approx$ €1,150,000
Maximum steep gradients	10%
Vertical acceleration (acceptable)	– 0.6 to + 1.2 m/s^2
Minimum planimetric radius (at maximum speed)	23.5 km
Minimum vertical radius (at maximum speed)	191 km

flow $q = v \cdot k = 1220 \cdot 0.026 = 31.87$ veh/h, minimum headway time between each pair of successive vehicles $\Delta t_{min} = 1/q = 1/31.87 = 0.0314$ h $= 113$ s and the theoretical system capacity $C = q \cdot P = 31.87 \cdot 28 = 892$ passengers/h.

Therefore, given the very small number of seats in each vehicle (28 seats) the practical transport capacity of hyperloop lines would only be around 892 passengers/h per direction. On the other hand, in the case of a bigger capsule with 40 seats, as those developed by Hyperloop Transportation Technologies in the year 2019, the estimated theoretical capacity is $C = 1.275$ passengers/h per direction.

It is worth underlining that in the case of mono-directional operation of a set of two parallel hyperloop tubes, train operations at stations could considerably increase the minimum headway time (Δt_{min}) to at least or more than 5 min [20]. Therefore the practical capacity of the hyperloop system can be estimated as $C = 3600/(5 \cdot 60) \cdot 28 = 336$ passengers/h per direction (far less capacity value than those of high-speed railway lines).

References

1. Pyrgidis CN (2016) Railway transportation systems. CRC Press
2. Profillidis VA (2014) Railway management and engineering. Ashgate
3. Cirillo B, Comastri P, Guida PL, Ventimiglia A (2009) High-speed railways (in Italian, L'alta velocità ferroviaria), Cifi
4. Campos J, de Rus G (2009) Some stylized facts about high-speed rail: a review of HSR experiences around the world. Transp Policy 16(1):19–28
5. Liu J (2010) Cost analysis of high speed railway and passenger dedicated line projects. J Railway Eng Cost Manage 25(3):20–23
6. Wu J, Nash C, Wang D (2014) Is high speed rail an appropriate solution to China's rail capacity problems? J Transp Geogr 40:100–111
7. Guerrieri M (2022) Hyperloop, HeliRail, Transrapid and high-speed rail systems. Technical characteristics and cost-benefit analyses. Research in Transportation Business and Management, vol 43. pp 100824
8. Han H-S, Kim D-S (2016) Magnetic levitation. Springer, Maglev Technology and Applications
9. https://railsystem.net/maglev/
10. Zhang RH et al (2004) Comparison of several projects of high speed Maglev. Adv Technol Electr Eng Energy 23(2):46–50
11. Xiangming W (2003) Maglev train. Shanghai Science and Technology Press, Shanghai
12. Almujibah H, Preston J (2018) The total social costs of constructing and operating a Maglev line using a case study of the Riyadh-Dammam corridor, Saudi Arabia. Transport Syst Technol 4(3):298–327
13. Schach R, Jehle P, Naumann R (2015) Transrapid und Rad-Schiene-Hochgeschwindigkeitsbahn. Springer
14. Wolters C (2008) Latest generation maglev vehicle TR09. Maglev 2008. San Diego, USA
15. Powell J et al. (2016) Maglev America how Maglev will transform the World economy. CreateSpace
16. Musk E (2013) Hyperloop alpha. SpaceX, Texas http://www.spacex.com/sites/spacex/files/hyperloop_alpha-20130812.pdf
17. van Goeverden K, Milakis D, Janic M, Konings R (2018) Analysis and modelling of performances of the HL (Hyperloop) transport system. Eur Transp Res Rev 10(2):41
18. https://vehiclecue.it/hyperloop-italia-bibop-gresta/17981/
19. Guerrieri M, Mauro R (2021) A concise introduction to traffic engineering. Springer
20. Hansen IA (2020) Hyperloop transport technology assessment and system analysis. Transp Plan Technol 43(8):803–820

Chapter 13
Metro Rail Systems

Abstract This chapter deals with the metro rail systems. The main technical characteristics of heavy metros and light metros are presented in terms of alignment (horizontal and vertical), grade of automation (GoA) and lines capacities.

Metro rail systems (also known as metros or underground railways)[1] are rapid mass transport systems, at high traffic capacity with high-frequency service (train headway up to 1 min) used in urban and suburban areas (see Table 13.1). With regard to the urban town centre, lines can have a radial direction (when they link the centre to suburban neighbourhood), a diametrical direction (when they pass through the centre) or tangential (when they develop around the centre).

The level of the travel demand identifies the system type to be used and therefore the metro rail systems can be classified into macro-categories:

- heavy metros
- light metros.

Heavy metros are often implemented in towns with over 1,000,000 inhabitants [1].

Light metros are so called for the lower capacity compared to heavy metros and the use of trains smaller in size and mass. They offer a capacity suitable to meet the transport demand in cities populated by around 500,000 and 1,000,000 inhabitants.

According to a survey updated to 2015 [1] there are metro rail systems in 155 world cities (59 in Europe, 61 in Asia, 33 in America, 2 in Africa) and they are being implemented in other 39 cities. The ten cities with the most developed lines are listed in Table 13.2.

Big cities may also provide mixed systems: heavy metros on lines with higher traffic demand and light metros on those with lower demand.

Metros can have a different grade of automation (GoA) (see Table 13.3) [2]:

- GoA1: manual operation; the driver is involved throughout the driving activities of the train, which is equipped with *Automatic Train Protection* (ATP) system;

---

[1] It was named after "the Metropolitan railway", the first underground railway open to the traffic in London on 10 January 1863.

© The Author(s), under exclusive license to Springer Nature Switzerland AG 2023
M. Guerrieri, *Fundamentals of Railway Design*, Springer Tracts in Civil Engineering,
https://doi.org/10.1007/978-3-031-24030-0_13

Table 13.1 Main technical characteristics of light and heavy metros [1]

	Automated light metro	Light metro	Heavy metro
Distance between successive stops [m]	400–800	400–800	500–1000
Commercial speed [km/h]	25–35	25–35	30–40
Grade separation	Underground or at grade	Underground or at grade	completely underground
Voltage [V]	750	750	750–1500
Power supply system	Catenary or ground-level ("third rail")	Catenary or ground-level ("third rail")	catenary or ground-level ("third rail") ("third rail")
Maximum capacity [passengers/hour/direction]	25,000	35,000	45,000
Train composition	2 coaches	2–4 coaches	4–10 coaches
Total train length [m]	26–27	60–90	70–150
Total train width [m]	1.62–1.88	2.10–2.65	2.60–3.20
Driving system	Automated	Manual or automated	Mainly automated

Table 13.2 Cities with the longest metro networks

Country	City	Inhabitants	Length of the metro network [km]
China	Shanghai	24,256,800	538
South Korea	Seoul	10,117,909	469
China	Beijing	21,516,000	465
United Kingdom	London	8,673,713	402
USA	New York	8,550,405	373
Russia	Moscow	12,330,126	325
Japan	Tokyo	15,185,502	317
Spain	Madrid	3,141,991	293
China	Guangzhou	12,700,800	240
Mexico	Mexico City	8,851,080	227

- GoA2: semi-automatic operation; the driver is involved only in case of system failure and is responsible for opening and closing the train doors. The train is equipped with *Automatic Train Operation* (ATO) and ATP systems;
- GoA3: totally automated operation; the train moves driverless but an attendant is responsible for opening and closing the doors and able to intervene in case of system failure. The train is equipped with ATO and ATP systems;

Table 13.3 Metro classification system based on the grade of automation in their operation [2]

Grade of automation	Train operation	Setting train in motion	Train driving and stopping	Door closure	Operation in event of failure
GoA1	Driver + ATP	Driver	Driver	Driver	Driver
GoA2	Driver + ATP + ATO	Automatic	Automatic	Driver	Driver
GoA3	Automated (attendant) + ATP + ATO	Automatic	Automatic	Attendant	Attendant
GoA4	Automated + ATP + ATO	Automatic	Automatic	Automatic	Automatic

- GoA4: totally automated operation; this system does not require either a driver or an attendant.[2] The train is equipped with ATO and ATP systems.

13.1 Heavy Metros: Horizontal and Vertical Alignment

The metro railway systems must be designed so that their alignment follows, with as little horizontal length as possible, the urban transport desire lines with higher traffic demand. Moreover, the influence zones of the stations must comprise the main production–attraction trip areas.

The alignment is designed in function of the characteristic speed, the value of which is 70–90 km/h [3]. The distance between two successive stations, together with the train operational speed, affects the commercial speed value which is usually around 30–40 km/h.

The geometric elements of the alignment must ensure that the maximum speed is at least equal to the assumed characteristic speed. The criteria for dimensioning the alignment are the same as those illustrated for ordinary railways (see Chap. 2) with the following further specifications.

The minimum radius (R_{min}) of horizontal circular curves must be:

- $R_{min} \geq 150$ m in line (but radii over 200 m are the most recommended);
- $R_{min} \geq 75$ m near turnouts and at sheds.

Between straight sections and circular curves, or between two circular curves, a transition curve of parabolic or clothoid type should be inserted.

If for technical reasons no reverse clothoid can be inserted between consecutive curves (with curvature of opposite sign), a straight section, $L \geq 50$ m long, can be interposed between the two reverse curves.

---

[2] This system is employed in several cities worldwide: https://en.wikipedia.org/wiki/List_of_auto mated_urban_metro_subway_systems .

The maximum rail superelevation (cant) in curve is set in 160 mm. The superelevation profile (developing along the parabolic or clothoid transition curves) must satisfy the relation p ≤ 3 mm/m (cf. Chap. 2).

The non-compensated acceleration (a_{nc}) must be lower than 0.4 m/s^2.

As regards the vertical alignment, the following values are necessary:

- grades with slopes preferably always below 30‰; in exceptional cases slopes can reach 50‰ [1].
- the vertical curves, obtained with arcs of a circular curve, must have a radius R_v ≥ 3.000 m.

13.2 Stations

Stations can be classified into:

- Transit stations;
- Terminal stations;
- Crossing stations.

Transit stations serve the areas with a higher travel demand concentration and, therefore, are sited near the production–attraction trip zones.

The influence area of a station can be assumed as a circle with radius R_s = 300–600 m within which potential users can find the transport service easy and suitable; within that area, in fact, the maximum walking time is inferior or equal to around 5 min[3] from the departure point (home, office etc.) to the station entrance. Thus, the theoretical distance between the two stations should be always inferior or equal to 2·R_s. Only for lines in suburban low-density areas the distance between stations can reach 2,000 m.

The commercial speed V_c is influenced from the average distance between the stations L_m. The following empirical relation between V_c (in km/h) and L_m (in metres) [4] is well known:

$$V_C = k \cdot \sqrt{L_m} \qquad (13.1)$$

where k = 1 for old stations and k = 1.41 for modern stations.

Based on the track position, in stations platforms can be lateral or central and island-shaped (cf. Chap. 8). The latter layout can imply a much wider distance between tracks than that usually employed in line sections, and reverse clothoid connections at the platform ends are required; therefore, this layout can be used exclusively in tracks placed into distinct tunnels, side by side and at a proper distance.

---

[3] In order to enlarge the influence area of a station, an entrance can be located at each of the two ends of the platforms. It has been shown that this layout, compared to the station with only one entrance sited near the centre line of the platform, increases the influence area of $\Delta A \approx 2 \cdot R_s \cdot L_b$, where L_b denotes the platform length [1]. For instance, if R_s = 500 m, L_b = 200 m, the surface increases as follows ΔA = 200,000 m^2.

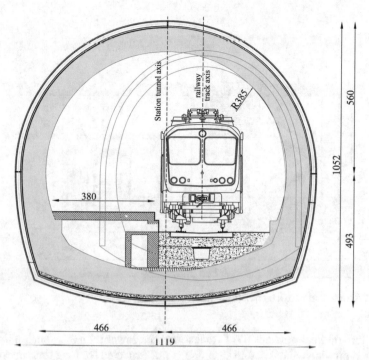

Fig. 13.1 Cross section of a tunnel at a station (dimensions in cm)

Platforms are 3.5–5.5 m wide (see Fig. 13.1); their length must be greater than that of the longest train expected in a metro. In order to avoid lift installation, the height difference between the metro platform and the road platform must be, where possible, limited. Staircases must have flights 1.40 m wide if travelled along one direction or 2.40–3.00 m wide if travelled on both directions (see Fig. 13.2). In any case, precise dimensions require controlling the expected pedestrian flows. Tread (t) and riser (r) must satisfy the empirical relation $2r + t = 62$–64 cm (e.g. riser 15 cm, tread 33 cm). Any moving staircases, besides the fixed ones, must be consistent (in terms of speed, inclination of flights, tread and riser of the steps etc.) with specific technical regulations [5].

The terminal stations allow travelling directions to be switched and/or train compositions to be changed (in number of coaches). They can be divided into dead-end stations and loop stations.

Finally, crossing stations have two or more levels and each level only serves one line. In these particular terminals, passengers can change line, running direction and forward direction.

For functionality and safety reasons, at stations tracks must be in straight sections and exceptionally in curves.[4]

---

[4] However all train doors must be visible from the driver's cab.

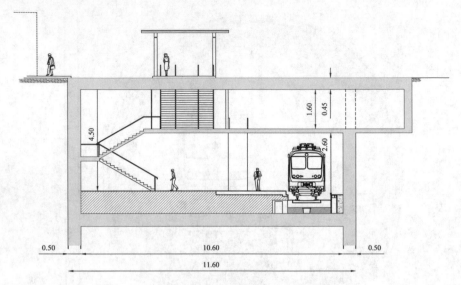

Fig. 13.2 Cross section of a metro station (dimensions in metres)

The longitudinal gradient (i) of tracks must be lower than 2‰; such a limit can be waived provided that the train stopping is not expected for long time intervals; in any case, it must be: i ≤ 20‰.

In order to make train braking and successive restart easier, at stations it is advisable that the arrival and departure grades show a value gradient which is, respectively, positive (ascent) and negative (descent) with a "cradle" vertical profile.

13.3   The Capacity

The theoretical capacity of a metro line, as occurs in other public transportation system types, is based firstly on the service frequency and the train capacity P, and can be estimated, only at a first approximation and in homotachic regime (cf. Chap. 11), with the relation:

$$C = \frac{3600}{\Delta t} \cdot P \tag{13.2}$$

In which C is the line capacity (expressed in passengers/hour per direction), Δt is the average headway between the train passage and the following (expressed in seconds) and P is the train capacity (passengers/train).

On account of the traffic management systems, station dwell times (15–30 s) and other factors, the minimum headway is Δt = 60–70 s but, frequently, not below 2–3 min.

Trains are composed of more coaches. Every coach is 15–20 m long, 2.60–3.00 m wide and with a carrying capacity of 160–250 passengers/coach and seating capacity percentage equal to around 30%. According to the train composition and considering that each train is generally composed of 4–6 coaches,[5] a number of people ranging from 600 to 1500 can be transported by each train.

By way of an example, if trains have a carrying capacity P = 1000 passengers/train and $\Delta t = 2$ min, from Eq. (13.2) it results: C = 3600/120·1000 = 30,000 passengers/hour for direction.

The maximum system capacity amounts to 45,000 passengers/hour for direction.

When the transport demand varies, the line capacity can be modified by acting on the frequency of transits and/or the composition of coaches; in any case the total length of a train must not be greater than the length of the platform.

13.4 Tunnels and Superstructure

Metro rail systems develop nearly entirely underground. As a rule, a metro network requires the construction of both artificial and natural tunnels (cf. Chap. 10). Unlike what occurs in natural tunnels, the line sections in artificial tunnels have a horizontal and vertical alignment bound by the above road network. In this case, the top surface of rails can be laid at a maximum depth of around 15 m below the road level. In order to allow underground services (underground pipes, cables and equipment associated with electricity, gas, water and telecommunications) to pass through, it is necessary to form an at least 1.50 m soil layer over the top slab of the tunnel. Compared to suburban railway tunnels, metros require further design analyses about, among other things, the most suitable excavation methods for limiting land subsidence and thus preserving building integrity, the study of vibration effects produced at the excavation stage and, more in general, the construction site activities which may interfere with those usually conducted in towns.

Among the most-widely used railway tracks, both the Stedef system (cf. Chap. 3) and the so-called "the massive Milan" system (floating slab track) have the advantage of mitigating vibrations and reducing maintenance interventions and related costs, in comparison with other superstructures.

13.5 Light and Automated Light Metros (VAL, Véhicule Automatique Léger)

According to some scholars, light metros can be considered hybrid systems between heavy metros and tramways [1].

---

[5] Also trains with two or more indivisible units are frequently used.

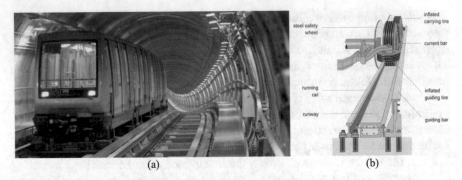

(a)   (b)

Fig. 13.3 **a** Automated light metro in Turin; **b** guide bar (*Source* [7])

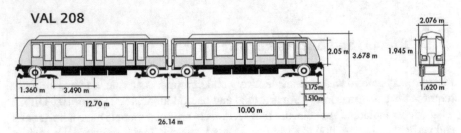

Fig. 13.4 Dimensions of VAL 208 train

Despite the lower capacity values (see Table 13.1), mainly due to small-sized trains (with rubber-tyred or steel wheels), light metros bear lower construction and management costs than the corresponding costs of heavy metros.

The automated light metros (Fig. 13.3a) employ small vehicles (length: 24–28 m, width: 2.1–2.9 m, height: 3.1–3.5 m, see Figs. 13.4, 13.5) equipped with rubber-tyred wheels, running[6] on guide bars (Fig. 13.3b) which also serve as power supply system ("third rail").

The vehicles have a carrying capacity of around 440 passengers (17% seats). Light metros are characterized by lower construction costs than conventional metros (around 50% less); the operational costs are also lower. The capacity is around 25,000 passenger/hour per direction; thus, they are transportation systems which can be employed in urban areas with such levels of demand that do not justify the implementation of heavy metros.

The basic characteristics of these transportation systems are [1, 9]:

- small vehicles equipped with an automated driving control system;
- tunnels with a small-sized cross-section which can correspond to 23–45% of the tunnel area of heavy metros, depending on the train type;
- short stations, for the limited length of platforms;
- gradient slopes (up to 13%) steeper than those suitable for heavy metros;

---

[6] The idea of building vehicles with rubber-tyred wheels was devised by Michelin in the 1950s [6].

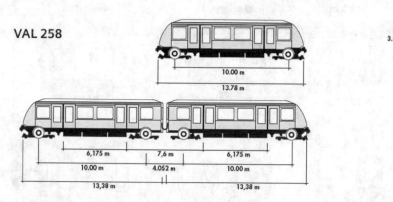

Fig. 13.5 Dimensions of VAL 258 train

- use of steering bogies and rubber-tyred wheels, which allow using small radius horizontal curves;
- vehicle floor at the same level as platforms (absence of steps);
- automated driving system does not require any on-board staff (absence of drivers);
- command and control system regarding all the transportation system units such as, for instance, train movement, switch operation, vehicle door opening, platform screen door opening and telecommunications;
- automated vehicle circulation (stop and door opening/closing at stops);
- power supply system inserted in the guide bars;
- vibrations and noises less than heavy metros;
- automation allowing the frequency of transit to be up to 60 s;
- weight of a vehicle pair equal to 31 tonnes for the VAL 208 (Fig. 13.4) and of 37 tonnes for VAL 258 (Fig. 13.5);
- acceleration and deceleration under normal conditions of motion equal to 1.3 m/s^2; emergency deceleration equal to 1.8–3.03 m/s^2.

The minimum horizontal radius R_{min} of circular curves is:

- $R_{min} = 30$ m for VAL 208;
- $R_{min} = 40$ m for VAL 258.

Automated light metros are supplied with direct current at 750 V. Every train is equipped with four engines, each with a power of 65 kW (260 kW total/train).

There are other automated transportation systems with characteristics similar to VAL metros like, for instance, the following (Fig. 13.6).

- Advanced Light Rail Transit (A.L.R.T.)
- Automatic Guideway System (A.G.S.)
- Automated Guideway Transit (A.G.T.).

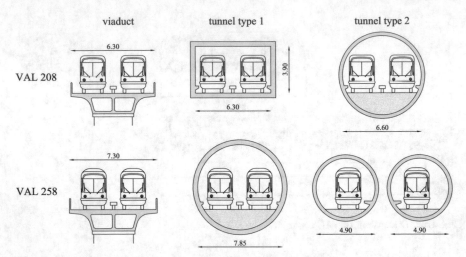

Fig. 13.6 Characteristic dimensions of viaducts and tunnels for VAL

13.6 Construction Costs

Construction costs depend on the vehicle type, grade of automation (GoA), line alignment, excavation methods, distance between stations etc. Therefore, the cost can range between € 60 million/km and € 130 million/km (M€/km) [1, 7]. In general, the following construction costs per km can be considered for heavy metros [1]:

- 70–90 M€/km, when the tunnel length is 75% of the total line length;
- 100–130 M€/km, when the line is wholly in the tunnel (100% of the total line length).

The train cost varies from 1.3 to 2 M€; thus, if for instance a train is composed of eight coaches, its cost ranges between 10.4 and 16 M€.

References

1. Pyrgidis CN (2021) Railway transportation systems. CRC Press
2. Bieberc CA (1986) Les choix techniques pour les transport collectifs. Ecole Nationale des Ponts et chaussèes, Paris
3. Policicchio F (2007) Railway infrastructures (in Italian. Firenze University Press, Lineamenti di infrastrutture Ferroviarie)
4. Tocchetti A (2008) Railway infrastructures (in Italian, *Infrastrutture Ferroviarie*). Aracne
5. D.M. 18///1975 Guidelines on moving staircases. (in Italian)
6. Nicolardi A (1956) Special railway systems (in Italian, *Ferrovie speciali*), Casa editrice Dott. Carlo Cya
7. https://railsystem.net/rubber-tyred-metro-2/
8. Profillidis VA (2022) In: Railway planning, management, and engineering. 5th edn. Routledge
9. Gargiulo A (2013) Mobility analyis in Rome (in Italian, *La mobilità a Roma*), Gangemi editore

Chapter 14
Tramway Systems

Abstract In this chapter the principal technical characteristics of tramways with conventional and ground-level power supply systems are briefly discussed. The various aspects influencing tramway alignments and superstructures are also examined.

Tramways are public mass transport infrastructures on which steel-wheeled electric trains move almost entirely at grade (and only exceptionally in tunnels or on viaducts) along roads in urban or suburban areas.[1] They can be built in exclusive separated corridors (i.e. in reserved lanes) or in common corridors (i.e. lanes shared with other traffic categories, including road vehicles, pedestrians, etc.). The main characteristics of the modern tramway systems and their infrastructures are:

- low-floor vehicles (no steps at doors for the ascent and descent);
- line extension average between 5 and 20 km;
- commercial speed of 12 –30 km/h;
- theoretical maximum capacity of around 6000 passengers/hour per direction.

The characteristic capacity values of the tramway system can be inferred by comparing other public transportation systems obtained on the basis of the headway (Δt) between vehicles. They are shown in Fig. 14.1.

14.1 Classification

Tramways can be classified based on:

- corridor type;
- service provided;
- superstructure employed.

---

[1] The first electrified tramway line was inaugurated on 16th May 1881 in Berlin. It was 2.45 km long (1 m gauge and 10‰ maximum gradient); the vehicles ran with a maximum speed of 15 km/h [1].

© The Author(s), under exclusive license to Springer Nature Switzerland AG 2023   239
M. Guerrieri, *Fundamentals of Railway Design*, Springer Tracts in Civil Engineering,
https://doi.org/10.1007/978-3-031-24030-0_14

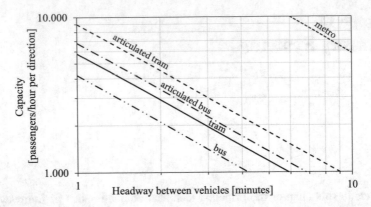

Fig. 14.1 Capacity of tramways and other public transportation systems in urban areas

14.1.1   Classification Based on the Corridor Type

According to Bieber [2, 3], tramways can be divided into (Table 14.1):

Table 14.1 Main technical characteristics of tramway lines

Gauge	Normal (1435 mm)
Minimum radius	R_{min} = 20–25 m (preferably R > 30 m)
Transition curves	Cubic parabola/clothoid (exceptionally absent)
Vehicle length	Single vehicle: 8–10 m, articulated 18–30, multiarticulated 25–45 m
Vehicle width	2.20–2.65 m
Floor	Low
Commercial speed V_c	Class E: V_c = 12–15 km/h Class D: V_c = 16 –18 km/h Class C: V_c = 18 –20 km/h Class B: V_c = 20 – 25 km/h Class A: $V_c \geq$ 30 km/h
Increase in V_c in case of intersection priority	15 –25% for classes D and B
Distance between stops	200–500 m
Maximum capacity	6000 passengers/hour per direction
Supply	600/750 V, cc
Rails	Class E: grooved Class D: grooved Class C: grooved Class B: grooved or Vignoles Class A: Vignoles (46 or 50 kg/m)

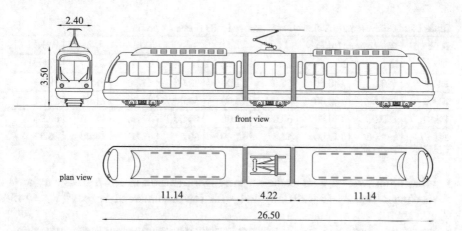

Fig. 14.2 Typical dimensions of a tram

- class E (*common corridor*): tram vehicles (Fig. 14.2) run on a lane of the carriageway together with other vehicle types (cars, motorcycles, bicycles etc.) and pedestrians. Trams run on grooved rails which are embedded in the road pavement. Their construction costs are relatively low but the commercial speeds are rather limited (12–15 km/h);
- class D (*exclusive separated corridor*): a road lane, equipped with grooved rails, is devoted to trams by means of horizontal traffic signs or curbs. The tramway lane may be occasionally run by other vehicles, thus causing delays to trams. The commercial speed is 16–18 km/h;
- class C (*exclusive tram corridor*): the whole carriageway is used as a tramway line, except lateral pedestrianised areas. It is a feasible solution only for sufficiently wide commercial roads or for historical city centres where private vehicles are not allowed to circulate;
- class B (*exclusive protected corridor*): the tramway line is physically separated from the remaining lanes of the carriageway by means of insurmountable curbs; alternatively, a railway track can be built on a reinforced concrete platform, which is superelevated over the road paving. The commercial speed is 20–25 km/h;
- class A (*fully exclusive corridor*): the tramway line does not exploit the existing road network but it develops along its own infrastructure (nearly always in suburban areas), as occurs for ordinary railways. The commercial speed exceeds 30 km/h.

14.1.2 Classification Based on the Service

Tramways are classified based on the type and extent of services provided as follows [2]:

Table 14.2 Characteristic dimensions of grooved rails (Phoenix profile)

Rail	Dimensions (mm)							Weight
	H	P	B	A	L	G	C	(kg/m)
Ri 59	180.00	180.00	56.00	12.00	42.00	47.00	15.00	58.19
Ri 60	180.00	180.00	56.00	12.00	36.00	47.00	21.00	59.74
Ph 37a	180.00	180.00	60.00	13.00	60.00	47.00	15.00	66.90
PH 37	182.00	150.00	52.5	11.00	60.5	46.00	14.00	56.50

- urban tramways: only for exclusive passenger transport in urban areas, on short-medium distances (5 –20 km) and with low commercial speeds (V_c = 15 –25 km/h);
- tram-train: thanks to specific technological devices,[2] vehicles can run along tramways and local railway lines (in urban and suburban areas) mixed with trains. The covered distances are of the order of 15 –50 km. The service is offered with the maximum speed V_{max} = 80 –120 km/h and commercial speed $V_c \approx$ 60 km/h;
- tourist tramways: these very short lines are characterised by extremely reduced commercial speeds and are devoted only to tourist traffic;
- tramways for freight transport: trains are only devoted to freight transport. They are not very widespread systems.[3]

14.1.3   The Superstructure

The superstructure (or railway track) employed for conventional tramways can be of two types: ballasted track (only in suburban areas) (Fig. 14.3) or ballastless track (see Chap. 3) (Fig. 14.4). The rails have nearly always a Phoenix[4] profile ("grooved", see Table 14.2).

14.1.4   Tramways with Ground-Level Power Supply System

In historical city centres tramways with ground-level power supply systems are more and more required, in that they have marginal environmental and visual impacts and prove to be adequate for the typical needs of that particular urban context.

Tramways with ground-level power supply systems, also called "catenary-free tramway systems" (see Fig. 14.5), bring clear benefits to sidewalks as well, since

---

[2] Trams are equipped with switching devices: they can operate as a railway (3 kV cc in Italy, 15/25 kV ca in the rest of Europe) and as a tramway (600/750 V, cc).

[3] This type of tramway operates in Zurich, Dresden and Cologne.

[4] The specific rolling resistance along Phoenix rails is greater than the resistance observed on Vignoles rails. According to [1] on tramway lines the wheel-rail adhesive coefficient (cf. Chap. 1) is supposed to have a value f = 0.13–0.14.

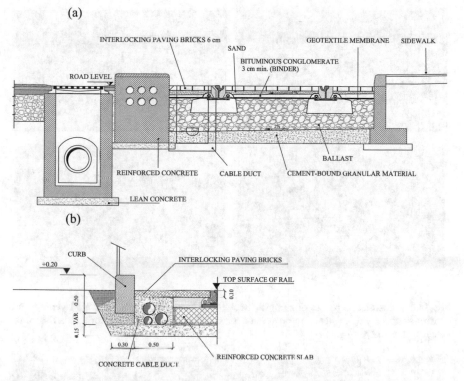

Fig. 14.3 Superstructure details: **a** ballasted track and **b** slab tracksfor class B tramways

a catenary-free tramway system does not require poles, hangers, conductors and brackets.

The following "catenary-free" technologies are currently available [5, 6]:

- *Ground-level power supply (GLPS)* (Figs. 14.5 and 14.6)—power continuously supplied to the vehicle at ground level via direct contact with a conductor or inductively. The most used system is Alstom APS, which may be found operating in the cities of Bordeaux, Angers, Reims, Orléans, Tours (France), Dubai (The United Arab Emirates), Rio de Janeiro (Brazil), Cuenca (Ecuador), Lusail (Qatar), Sydney (Australia).

   An innovative technology is the induction system (e.g. Primove Bombardier, Fig. 14.6b) which operates by the physical principle of electromagnetic induction;
- *Onboard energy storage system (OESS)*—power stored on the vehicle, using flywheels, batteries (Ni–MH; Li–Ion, etc.), supercapacitors or a combination thereof, recharged periodically via regenerative braking and contact with a power conductor. The autonomy of each vehicle is of about 600 m. The stops need to be equipped with recharging systems;
- *Onboard power generation system (OPGS)*—power continuously generated on the vehicle as required via hydrogen fuel cells, micro-turbines or diesel engines.

(a)   (b)

(c)   (d)

Fig. 14.4 Construction phases of a railway track [4]: **a** casting of a lean concrete layer and curb construction; **b** laying of steel reinforcement bars; **c** laying of the flat framework; **d** casting of the concrete slab and paving

Fig. 14.5 *GLPS*: TramWave system by Ansaldo STS[5]

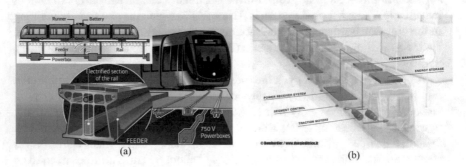

(a)   (b)

Fig. 14.6 *GLPS*: **a** APS Alstom (*Source* www.alstom.com); **b** Primove Bombardier—wireless (*Source* www.bombardier.com)

Fig. 14.7 Rubber-tyred tram (Translohr)

14.1.5 The Rubber-Tyred Tram

Rubber-tyred trams are particularly interesting vehicles, in that their wheels are equipped with tyres and the track has only one rail.

Among the most well-known systems, there are [7]:

- Guided Light Transit (GLT);
- Translohr (Fig. 14.7).

Both systems make use of small metallic guiding wheels but with different geometry (90° for GLT and 45° for Translohr).

The Translohr[6] system allows for designing alignments with horizontal radii even of modest value, since trams have a minimum steering radius of 10.5 m; moreover, the maximum longitudinal gradient assumes the value of 13%.

Vibrations and noise are markedly lower than conventional systems and also the implementation costs are usually more limited. On the other hand, trams run on a lane of the carriageway along with other vehicles; thus, safety problems for cyclists and motorcyclists have arisen for the peculiar type of rails used in this system.

Among the rubber-tyred systems there are the Autonomous Rail Rapid Transit (ART) with "optical driving" and trackless infrastructure (Civis-type). The ART route is led by optical sensors which are able to detect the virtual track obtained by road signs (broken double white lines, see Fig. 14.8) and therefore the trajectory of the tramway line. A three-carriage ART train is about 30 m long, its carrying capacity is of around 300 passengers and its maximum speed can reach 70 km/h.

New technologies are being experimented to allow platoons composed of a convoy of automated driving vehicles to run (intervehicle headways up to 1 m), by analogy

---

[5] http://www.ansaldo-sts.com.

[6] Today lines in operation are in Clermont-Ferrand, Medellin, Mestre-Venice, Padua, Shanghai, Tianjin, Paris.

Fig. 14.8 ART system proposed by the CRRC Corp. Ltd in 2017

to the platooning principle of Automated Highway System (AHS[7]) and smart roads
[7, 8].

14.1.6   Road Safety Analyses

In order to identify the risk factors associated to tramways [9–16] and the technical
countermeasures which need to be adopted to safeguard road users, it is advisable
to include specific procedures on safety analysis, like the Road Safety Audit (RSA)
in the design phase. This type of audit aims at identifying and correcting the technical choices made, for instance, for the tramway alignment and the traffic priority
systems which may be risky to all the users' categories (drivers, pedestrians, cyclists,
passengers etc.). The RSA can be structured by analogy to the guidelines given in
[16].

14.1.7   Construction Costs

The average construction cost of conventional tramways is 20 –23.5 M€/km
(20 M€/km in Africa, 22.5 M€/km in Europe, 23.5 M€/km in North America).

The construction cost of tramways with ground-level power supply system is
about 20% higher than that of conventional tramways.

The cost of each tram varies between 2.5 and 3.5 M€. On the other hand, the cost
of a train for a tram-train system is 4–4.5 M€ [2].

---

[7] In the AHS the traffic flow is composed of automated or semi-automated driving vehicles, travelling
as individual units or in platoons with potential benefits in terms of capacity and safety.

References

1. Nicolardi A (1956) Special railway systems (in Italian, *Ferrovie speciali*). Casa editrice Dott. Carlo Cya.
2. Pyrgidis CN (2021) Railway transportation systems. CRC Press, 2016
3. Bieberc CA (1986) Les choix techniques pour les transport collectifs. Ecole Nationale des Ponts et chaussèes, Paris
4. Tramway lines in Palermo. CIFI Symposium, 12 May 2010
5. Vuchic VR (2007) Urban transit systems and technology. Wiley
6. Guerrieri M (2019) Catenary-free tramway systems: functional and cost-benefit analysis for a metropolitan area urban rail. Transit 5(4):289–309
7. Guerrieri M, Ticali D (2012) Sustainable mobility in park areas: the potential offered by guided transport systems. In: ICSDC 2011: integrating sustainability practices in the construction industry—proceedings of the international conference on sustainable design and construction 2011, pp 661–668
8. Évolution des réseaux de transport urbain guides Revue Générale des Chemins de Fer. (1998) 1998(2):33–40, 61
9. Guerrieri M (2021) Smart roads geometric design criteria and capacity estimation based on av and cav emerging technologies. Trans-Eur Transp Netw Int J Intell Transp Syst Res 19(2):429–440
10. Ioannou PA et al (1997) Automated highway systems. Springer
11. Fontaine L, Novales M, Bertrand D, Teixeira M (2016) Safety and operation of tramways in interaction with public space. Transp Res Procedia 14:1114–1123
12. Korve HW et al (1996) TCRP report 17: integration of light rail transit into city streets. Transportation Research Board, National Research Council, Washington, D.C.
13. Korve HW et al (2001) TCRP report 69: light rail service: pedestrian and vehicular safety. Transportation Research Board, National Research Council, Washington, D.C.
14. Pecheux KK et al (2009) TCRP synthesis 79: light rail vehicle collisions with vehicles at signalized intersections. A Synthesis of Transit Practice, Transportation Research Board, Washington, D.C.
15. Guerrieri M (2018) Tramways in urban areas: an overview on safety at road intersections. Urban Rail Transit 4(4):223–233
16. Road safety audit guidelines. FHWA, 2006

Chapter 15
People Movers, Monorails and Rack Railways

Abstract This chapter deals briefly with the principal technical characteristics of people movers, monorails and rack railways. It also examines the various aspects affecting horizontal alignments, vertical alignments and superstructures of such transport systems.

15.1 Automated People Movers (APMs)

People movers[1] (or Automated People Movers—APMs) are public mass transportation systems, automated, with a dedicated steel or concrete guideway (generally elevated), usually with medium-sized two-axle electric vehicles operated automatically and with no crew on board [1]. They are suitable for single vehicles with capacity of 3–25 people or trains with 2–6 carriages (Fig. 15.1) with the total carrying capacity of 50–250 (sitting or standing) people.

The transport system capacity can reach up to 8000 passengers/hour per direction [1, 2]; the maximum speed—according to the utilised vehicle type—is 35–70 km/h.

The total length of lines is mostly comprised between 300 and 12,000 m, therefore some authors consider APMs as a hectometric transportation system[2] [3].

Depending on their propulsion system type, APMs can be classified into:

- self-propelled vehicles: they have an autonomous propulsion, being equipped with an onboard engine. The adopted electrification systems are [4]:

  - direct current 750 or 1500 V;
  - alternating current 480 or 600 V.

  The current is provided by a power distribution subsystem along the guideway.

- cable-propelled vehicles: they do not move autonomously; the movement is transmitted from a fixed electric motor, placed at one end of the line, by means of a

---

[1] The following terms are used without distinction: Automated Guided Transit (AGT), People Mover Systems (PMS), Group Rapid Transit (GRT), Personal Rapid Transit (PRT) [1].

[2] It is a transportation system in which the total length of the lines ranges from a hectometre (100 m) to 20 hectometres (2000 m) approximately [3].

Fig. 15.1 People mover: a line and a station

haul rope to which vehicles are coupled. In this case, the electrification system employed is the direct current at 480 or 600 V [4].

Depending on the line gradient, APMs can be divided into [5]:

- systems suitable to overcome elevated differences in height (longitudinal gradient $i \geq 10$–15%);
- systems suitable to overcome medium-low differences in height (longitudinal gradient $i < 10$–15%).

The vehicles have nearly always rubber-tyred wheels and very rarely steel wheels; only in a few cases they are endowed with a magnetic levitation propulsion system [4] and run on specifically designed guideways in function of the train technology.

On lines of medium-small lengths up to 2–3 km, the propulsion system is nearly always made of one or more pulling cables (guidance sheaves) with compacted strands running on the elevated guideway (Fig. 15.2), while the vehicle can be provided with a self-propelled system on longer lines.

The APM stations can be of two types: unattended stations (thus requiring a surveillance system linked to a control centre) or temporary unattended stations (e.g. at specific time of day or days of the week).

Fig. 15.2 Details of guideways with guidance sheaves. *Source* [5]

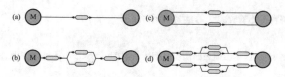

Fig. 15.3 APMs with cable-propelled systems and shuttle-based configuration: **a** single-lane shuttle; **b** single-lane shuttle with bypass; **c** double-lane shuttle; **d** double-lane shuttle with bypass

Fig. 15.4 Loop configurations: **a** single loop, **b** double loop and **c** pinched loop

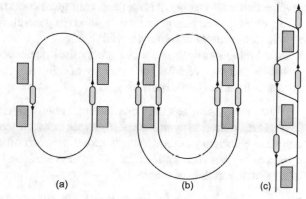

APMs are typically employed for connecting an airport terminal[3] to the city centre (when the distance is relatively short), even if there are some APM lines aiming at exclusively meeting the traffic demand of specific urban areas or university campuses [4]. The service takes place [4, 5]:

- *with a shuttle-based configuration*: a vehicle or a convoy of vehicles (train) of big capacity (generally, up to 250 passengers) is permanently fixed to the pulling cable (non-detachable system) and shuttles between two or more stations (Fig. 15.3). In order to improve the system capacity, an APM employs two trains, running among the terminal stations, along a single guideway with a bypass area (one or two pulling cables can be used);
- *with a continuously circulating configuration*: small-sized vehicles (up to 50 passengers) are not firmly anchored to the pulling cable and come uncoupled near stations (detachable system) which activate the deceleration system. There are single-loop, double-loop and pinched-loop systems (Fig. 15.4).

---

[3] In 2005 these transportation systems were already in operation in 26 world airports with lines from 700 m (Pittsburgh) up to 10 km (San Francisco) and with most frequent lengths ranging between 1 and 4 km.

15.1.1   Shuttle-Based Configuration

For the shuttle-based configuration the following configurations can be set up (Fig. 15.3) [5]:

- *type a (single-lane shuttle):* the system works with a pulling cable, one vehicle (or a convoy of vehicles) runs along a single track in conjunction with the two terminals (when necessary, some intermediate stops can be planned); the travelling direction is alternating. It is the same configuration as for funiculars (characterised by very elevated line gradients) or, more in general, for the APMs used in areas with lower levels of traffic demand;
- *type b (single-lane shuttle with bypass)*: the system works with one or two pulling cables and two vehicles (or a convoy of vehicles) running along a single track which links the two terminals.

When the system has only one pulling cable, vehicles cross each other near a bypass area placed at symmetrical distances between terminals. On the other hand, a system with two pulling cables is essential when differently long interdistances are necessary between stations (i.e. the bypass area can be set on a different section other than the middle of the line);

- *type c (dual-lane shuttle)*: the system works with two separate tracks, two vehicles and two pulling cables. A number of intermediate stations (generally, no more than three) can be implemented. In off-peak hours only one track can be used—like in type a), thus halving the system capacity;
- *type d (dual-lane shuttle with bypass)*: the system works with two tracks with an intermediate bypass and four trains. This configuration provides higher capacity and better service levels. In off-peak hours only one track can be used with a service analogous to type b), described above.

The capacity of an APM with a shuttle-based configuration can reach up 6000 passengers/hour per direction. This configuration is generally used in cases of small line lengths (up to 3 km) [5].

15.1.2   Loop Configuration

In this configuration, transportation vehicles are uncoupled from the pulling cable near stations that may be quite numerous. The number of operational vehicles depends on the traffic demand: few vehicles with low demand values and a great many vehicles with elevated demand values.

Varied configurations can be hypothesised (Fig. 15.4):

- *single loop*: the movement of vehicles takes place in one direction only, therefore users are obliged to travel along the ring section between the origin station and the destination station, also in case of long routes;

- *double-loop*: the vehicles run in both directions, therefore some routes are less than in the previous configuration. The capacity is twice that of the single loop and, in case of failure or maintenance on one loop, the other can be kept in operation;
- *pinched loop*: there are two tracks, like in the double-loop; however, in this configuration all vehicles use both tracks, which are linked with automated switches sited near the stops.

15.1.3 Guideway

The guideway is made of steel or steel and concrete. The line is nearly always superelevated for the whole extension so as to avoid interferences with the road network. The height of the guideway compared to the underlying road level must be at least equal to 5 m [6].

15.1.4 Construction Costs of APMs

The construction cost of APMs varies from 12.7 to 26.5 M€/km [5].

15.2 Funiculars

Funicular railways can be considered special types of APM systems. They are characterised from a limited length of lines, which usually have a horizontal length lower than 1.2 km, although some funiculars reach 5 km, and from the elevated longitudinal gradient with a constant value (300–500‰ and above) (Table 15.1).

The alternating movement service (i.e. shuttle-based configuration, see Fig. 15.3) is operated with two vehicles permanently linked to the pulling cable.

The line speed is strongly influenced by the longitudinal gradient and, more often than not, reaches lower values than 20 km/h.[4]

The lane can be single-track with two or three rails, or double-track with four rails (see Figs. 15.5 and 15.6). When the funicular is used for material transport, it is called "inclined elevator" [7].

---

[4] All over the world around 250 funiculars are in operation; train speed varies from 3.6 km/h to 50.4 km/h.

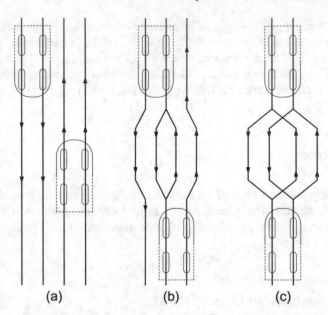

(a)   (b)   (c)

Fig. 15.5 a Lane with double track and four rails, **b** lane with single track and three rails; **c** lane with single track and two rails

Fig. 15.6 Alpenzoo terminal (designed by Zaha Hadid), funicular in Innsbruck ($i_{max} = 470\ \text{‰}$)

Table 15.1 Main technical characteristics of funicular systems

Maximum and minimum length of funicular railways in the world [m]	$L \leq 1200$ m (in general); $L_{min} = 39$ m, $L_{max} = 4827$ m
Gradient	$i = 300\text{–}500\%o$, $i_{min} = 90\%o$, $i_{max} = 1200\%o$
Minimum radius in horizontal circular curves	$R_{min} = 85\text{–}90$ m
Speed	$V \leq 20$ km/h
Most used gauge	1000 mm
Superstructure type	Single-track (2 or 3 rails) or double-track (4 rails)
Carriage number	2
Movement	Alternating (shuttle-based configuration)
Vehicle carrying capacity	10–420 passengers (more frequently 50–80)

15.3 Monorails

They are systems for passenger or rarely for freight transport in which a train, composed of 2–6 vehicles, runs above its guideway (straddled system) or below it (suspended system). The guideway is composed of one beam which has structural and driving functions, termed "guidebeam" (Figs. 15.7 and 15.8).

Monorails are designed for very short routes from 1.5 to 12 km. The maximum train speeds are 60–90 km/h, the commercial speeds are 20–40 km/h and the accelerations and decelerations reach $1.0\text{–}1.2$ m/s^2.

Table 15.2 Main technical characteristics of monorail systems

Maximum and minimum length of monorails in the world	$L_{min} = 1500$ m, $L_{max} = 12{,}000$ m
Maximum gradient i	$i_{max} = 100\%o$, exceptionally $i_{max} = 200\%o$
Minimum radius of circular curves for "light systems"	$R_{min} = 40$ m
Minimum radius of circular curves for "heavy systems"	$R_{min} = 70$ m
Minimum vertical radius	$R_v = 500$ m
Maximum speed	$V_{max} = 60\text{–}90$ km/h
Commercial speed	$V_c = 20\text{–}40$ km/h
Guidebeam width for light/heavy systems	$B = 2.30\text{–}2.64/3.00$ m
Capacity	12,500 passengers/hour per direction
Distance between the stops	800–1500 m
Electrification systems	750 cc–1500 V, cc

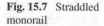

Fig. 15.7 Straddled
monorail

Given the particular and uncommon point of view presented to passengers, this transport system is frequently adopted in park areas and zoos but also in urban centres of big cities since its working is freed from the road traffic.

Based on the configuration, they can be classified into [8]:

- *straddled monorails*: train support and traction are obtained with wheels, usually tyred, some of which roll above the beam (bearing wheels) while others roll along the sides of the beam (driving wheels).

The guidebeam has a width of 670–700 mm for lighter systems with modest-sized vehicles, which are 2.30–2.64 m wide and have an axial weight of around 8 t (capacity of the order of 2000 passengers/hour per direction); while it has a width of 850–900 mm for heavier systems with vehicles which have a width of 3 m and an axial weight of around 11 t (capacity of 12,500 passengers/hour per direction) (Table 15.2).

The propulsion nearly always occurs with electric motors supplied with direct current;

- *suspended monorails*: the vehicle moves below the driving beam, to which it is properly bound;
- *cantilevered monorails*: the vehicle is suspended laterally to the beam.

In urban areas the lines are two monorails (a monorail for each direction) while in park areas where the traffic levels are relatively modest, only one monorail can be built with continuous movement service (loop configuration, see Fig. 15.4).

15.3.1 Monorail Construction Costs

For every monorail the unit cost varies from 30 to 90 M€/km depending on the configuration type.

Fig. 15.8 Suspended monorail

15.4 Rack Railways

In conventional railways, very elevated gradients (greater than 50–70‰) can be overcome by introducing a rack which is a linear system made of a toothed rack rail in-between the rails of a track. One or more cog driving wheels (that are positioned either horizontally or vertically in particular types of trains qualified for rack lines) are meshed with the toothed rack rail.

Cog rails are only installed on limited sections of ordinary lines, where the gradient overcomes 50–70‰, even if there are cases of lines equipped with rack systems for their entire length.

The most used cog systems are Locher, Strub, Riggenbach and Abt (Fig. 15.9). Among the current 65 rack lines operating in the world, the most used gauge is 1000 mm (over 50% of lines), followed by the standard gauge equal to 1435 mm and by the narrow gauge of 800 mm [5] (Table 15.3). Because of the drastic reduction in maximum and commercial speeds, which are more and more decreasing when the longitudinal line gradient increases (Fig. 15.10), rack railways have rather exclusively a tourist function nowadays [10, 11].

(a)   (b)   (c)   (d)

Fig. 15.9 Cog systems: **a** Locher, **b** Strub, **c** Riggenbach, **d** Abt

Fig. 15.10 Maximum speed downwards for rack railways in function of the longitudinal slope (data from [9])

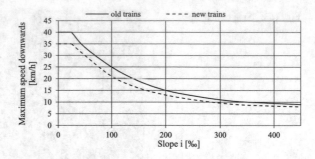

Table 15.3 Main technical characteristics of rack railways

Maximum and minimum length of rack railways in the world	$L_{min} = 1136$ m, $L_{max} = 19{,}090$ m
Maximum slope	$i_{max} = 250‰$ (exceptionally 480‰)
Minimum horizontal circular curve radius	$R_{min} = 90$ m
Minimum vertical curve radius	$R_v = 500$ m
Gauge	1435, 1000, 800 mm
Maximum speed	$V_{max} = 40$ km/h
Commercial speed	$V_c = 7.5$–20 km/h
Maximum acceleration	0.25–0.30 m/s^2
Train carrying capacity	100–200 passengers

15.4.1 Construction Costs of Rack Railways

The cost of a single-track line ranges between 10 and 15 M€/km [5].

References

1. Vuchic VR (2007) Urban transit systems and technology. Wiley
2. Commission staff working document impact assessment. Commission legislative proposal for a revision of Directive 2000/9/EC relating to cableway installations designed to carry persons
3. Corriere F (2011) Hectometric transportation systems (in Italian, Impianti ettometrici e infrastrutture puntuali per i trasporti). Franco Angeli Editore
4. Lea Elliott et al (2010) Guidebook for planning and implementing automated people mover systems at airports. ACRP Report 37. TRB
5. Pyrgidis CN (2021) Railway transportation systems. CRC Press
6. IMIT (2001) Guidelines for the design of road infrastructures: D.M. n. 6792, 5/11/2001. Italian Ministry of Infrastructures and Transports, Rome, Italy (in Italian)
7. Nicolardi A (1956) Special railway systems (in Italian, Ferrovie speciali), Casa editrice Dott. Carlo Cya
8. Liu R (2017) Automated transit. Planning, operation, and applications. Wiley

9. Guerrieri M, Ticali D (2012) Sustainable mobility in park areas: the potential offered by guided transport systems. In: ICSDC 2011: integrating sustainability practices in the construction industry—proceedings of the international conference on sustainable design and construction 2011, pp 661–668

10. Guerrieri M, Ticali D (2012) Design standards for converting unused railway lines into greenways. In: ICSDC 2011: integrating sustainability practices in the construction industry—proceedings of the international conference on sustainable design and construction 2011, pp 654–660

11. Loosli H (1984) Le chemin de fer à crèmaillère—ses particularitès et domaines d'application. Revue Technique Sulzer 2:17–20

Index

Printed in the United States
by Baker & Taylor Publisher Services